DIESEL MODEL ENGINES

DIESEL MODEL ENGINES

By

LT.-COL. C. E. BOWDEN, A.I.MECH.E.

By the same author
" Petrol Engined Model Aircraft "
" Model Jet Reaction Engines "
" Model Yacht Construction and Sailing "
" Model Glow Plug Engines "

First published 1947
Second impression 1948
Revised edition 1949
Revised edition 1951
Reprinted 1993
 ISBN 1 85761 035 0

WARNING

It should be remembered that the materials and practices described in this publication are from an earlier age when we were less safety conscious. Neither the methods nor materials have been tested to today's standard and are consequently not endorsed by the publishers. Safety is your responsibility and care must be exercised at all times.

CONTENTS

FOREWORD

When I wrote the first edition of this book the model diesel was in its infancy. The speed at which the book sold has been an indication of the very great interest in the diesel engine. Diesels are seen on every flying field and at every pond. During the past few years the diesel has made great technical strides in development in Britain, becoming the most popular type of power unit for model aeroplanes and boats, whilst the car enthusiasts are now embracing the type.

Low cost and low weight, allied to the lack of complication, has been the secret of the diesel's rapid success, and brought power modelling within the range of vast numbers who held back before its advent.

In America the growth of the diesel has been far less spectacular, for American modellers seem to favour very high r.p.m. and often large cubic capacity. As a result the existing petrol motor has developed along the lines of the glow plug engine which, like the diesel, eliminates the necessity of carrying weighty ignition gear out of all proportion to the weight of the motor. The glow plug engine thrives on high r.p.m., whereas the diesel, generally speaking, develops its power at slightly lower speeds with good torque characteristics. A very great advance in power output was made during the year 1949, as the result of control line flying influence and demands. Radio controlled models are now encouraging the larger capacity diesel.

This book is designed to put the newcomer to power modelling into the picture of the diesel. It explains how it works and the snags that many people may experience; and I hope will iron out any difficulties of starting, fuel, and operation, besides giving a survey of suitable models for diesel engines. Modern diesel engines are also reviewed.

I want to thank the following publications for kindly permitting me to use certain material—The *Aeromodeller*, *Model Aircraft*, *Model Cars*, *The Model Car News*, *The Model Engineer*, *Practical Mechanics*.

Bournemouth, 1951. C. E. BOWDEN.

CHAPTER I

HOW THE MODEL DIESEL WORKS, ITS DESIGN AND INSTALLATION

THE INTERNAL COMBUSTION ENGINE AND THE DIESEL

THERE is a good deal of misconception concerning the model diesel engine. For instance, many people ask how the injector mechanism operates, although in fact there is no such mechanism fitted to a normal model diesel.

It is therefore not out of place, bearing in mind that this book is written to help the complete novice as well as the more knowledgeable man, to describe at the outset the model diesel in elementary terms.

The model diesel as we know it at the time of writing this book is a development of an offshoot from the two-stroke model petrol engine, which, of course, is an " internal combustion " engine, i.e. it derives its power from a series of expansions of gas. The gas is created from liquid fuel which is atomised and mixed with air in the " carburettor." The expansions of gas are usually termed " explosions " by the general public. It is as well to explain that the word diesel is not strictly correct. It is the popular term for a " compression ignition engine," or " C.I. engine." A Dr. Diesel originally developed the C.I. engine, hence the well-known term " Diesel engine," which is now universally used in the same way that " propeller " is universally used to describe a tractor airscrew. These terms are so universal that they are used in official Service textbooks. Let us therefore call it the diesel engine in this handbook. The name is simple, and it suits everyone except the pedant.

To return to our expansion of gases or explosions in the internal combustion engine; the model internal combustion engine, like its full-sized prototype, can be a " four-stroke " or a " two-stroke." A four-stroke usually has poppet valves and fires on every fourth stroke of the piston. There are very few commercial four-stroke engines at present for model work. The two-stroke is almost universally sold for this purpose, because of its simplicity of construction. The normal two-stroke eliminates mechanically-operated valve gear (except in certain very advanced designs of full-sized engines) and it fires once every revolution of the crankshaft, i.e. each time the engine turns one complete revolution, or every second stroke of the piston.

IGNITING THE GASES

The petrol two-stroke engine fires its charge of gas by an electrical spark produced by means of a sparking plug when the piston is at the top of its stroke and has compressed the gas that it has sucked in.

The model diesel, at present, may be loosely likened to a petrol engine two-stroke without the electrical spark gear. It also fires the charge of gas when the piston is at the top of its stroke and has compressed the gas, as in the case of the petrol engine. Now, why does it do this when there is no sparking plug and no spark?

That is the crux of the whole matter.

The model petrol engine compresses its charge of gas to approximately one-fifth or one-sixth its original volume. The gas is compressed into a small space in the cylinder head as the piston comes up to the top of its stroke. This gas is then ready to " explode " *if it is ignited.* The electrical spark from the sparking plug produces this ignition and explosion follows. The piston is then driven down and the engine turns round. This action is shown in Fig. 1 and Fig. 2.

The model diesel engine compresses its gas in the same manner, but to approximately one-sixteenth the original volume, when the piston arrives at the top of its stroke. This top of the stroke is called " Top dead centre," abbreviated to T.D.C.

Now you well know that if you rapidly pump up your bicycle tyre, the quick compression of the air makes the bicycle valve and the bottom of the pump hot. This is because the air itself becomes hot due to the rapid compression and friction of the particles of the air. The diesel compresses the air it draws in even more rapidly, and at a greater pressure. In fact, it becomes so hot that if there is a suitable fuel mixed with the air it " explodes " from the heat *without a spark.*

In the full-sized diesel, pure air only is compressed by the piston and, as the piston arrives at the top of its stroke, a small carefully metered amount of diesel fuel oil is shot through a very tiny nozzle into the very hot compressed air. This forms a

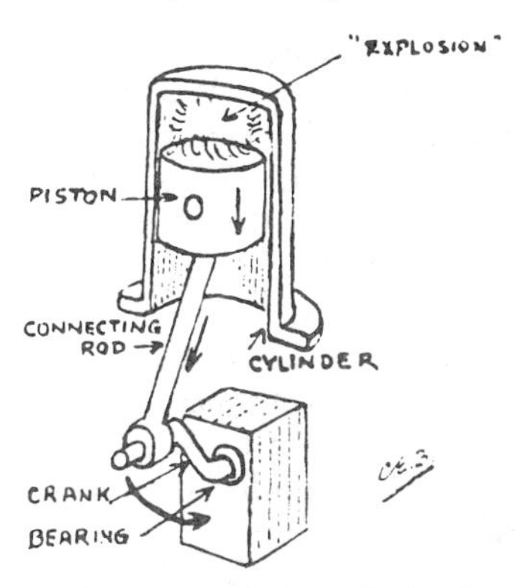

Fig. 1. The first principles of the internal combustion engine.

gas and explodes. The fuel injector gear to shoot the fuel into the cylinder against great pressure is complicated and expensive to produce, and far too heavy for a model; in addition, the amount of fuel to be injected in a model engine is so small that few craftsmen could make so small a pump. To overcome this, some clever individuals on the Continent thought of the idea of increasing the ignitability point of diesel oil and other fuels by adding ether to them. The model engine then sucked in its charge of air together *with the fuel containing ether*, and when the piston compressed this air and fuel mixture at its very high " compression ratio " of 16 to 1, the mixture became so hot as it arrived at the top of the stroke (T.D.C.) that it " exploded " *without any spark.*

It would obviously not be an economical practice for a full-sized diesel to burn ether because of the expense. But the model uses very little, and the cost does not enter the picture. However, it is interesting to note that in Czechoslovakia there is a little diesel scooter that runs on an ether mixture.

I need not explain that the compression must be exactly right to get the mixture at the correct temperature to suit the "flashpoint" of the fuel used. "Flashpoint" means the point at which a fuel ignites spontaneously, and ether ignites at a very much lower point than diesel oil, petrol, paraffin and other distillations of crude oil. Hence it has what is called a "low flash point," and "great ignitability."

It will also be appreciated at once that if the compression ratio is to be raised to more than double that of a petrol engine, i.e. from approximately 6 to 1 to 16 to 1, there will be vastly increased strains and stresses in the whole engine structure, therefore, although the diesel may be described as similar to a petrol two-stroke engine with increased compression, it *must be more robustly designed.* When I give 16 to 1 as the ratio, this is only approximate, as diesels fire on different ratios from about 12 to 20 to 1.

ADVANTAGES OF THE DIESEL

What advantage do we get from using a diesel instead of a petrol engine, because we know that we have the considerable disadvantage of these higher stresses and strains due to the very much higher compression ratio?

There is the elimination of all the trouble of electrical ignition gear and therefore of producing an ignition coil, a condenser, make and break gear, and sparking plug; and the trouble of wiring; cleaning and adjusting of the gear, etc., goes by the board. There is the saving of the weight of this gear, which in the case of the really midget petrol engine means a great deal, because the weight of the ignition gear on a petrol engine cannot be reduced below a certain minimum figure that is entirely out of proportion to the weight of the engine in the very minute sizes. The larger petrol engine scores in this

Fig. 2

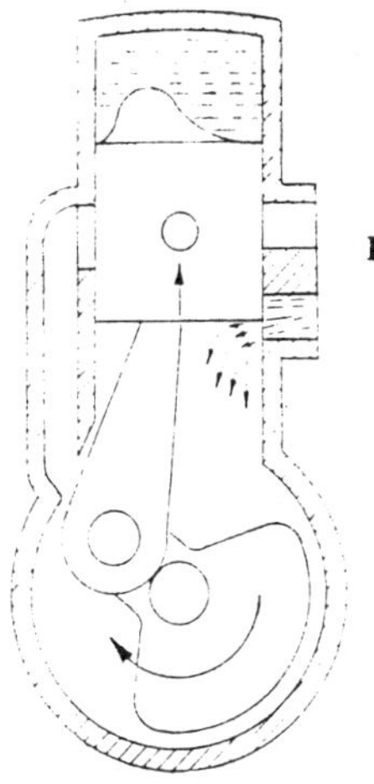

PISTON RISES SUCKS GAS INTO CRANKCASE COMPRESSES GAS IN CYLINDER

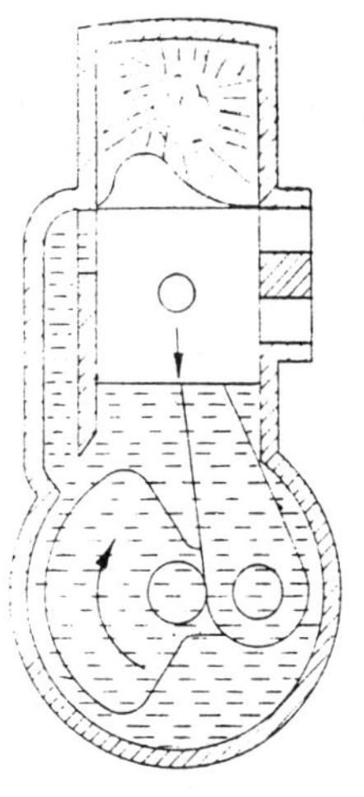

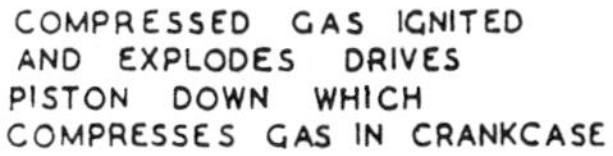

COMPRESSED GAS IGNITED AND EXPLODES DRIVES PISTON DOWN WHICH COMPRESSES GAS IN CRANKCASE

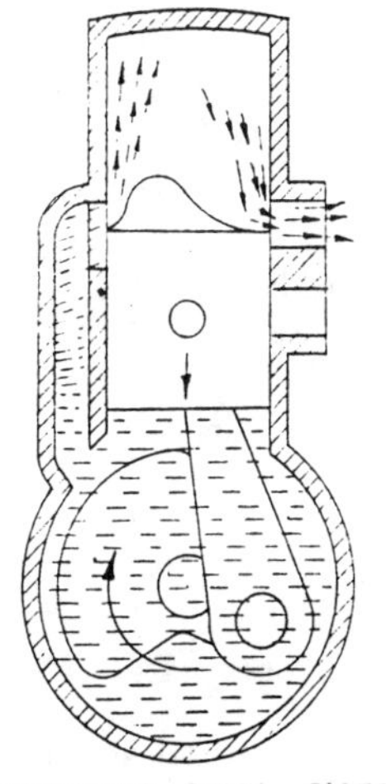

CONTINUING DOWN PISTON UNCOVERS EXHAUST PORT RELEASES BURNT GAS TRANSFER PORT OPENS CRANKCASE GAS RUSHES INTO CYLINDER

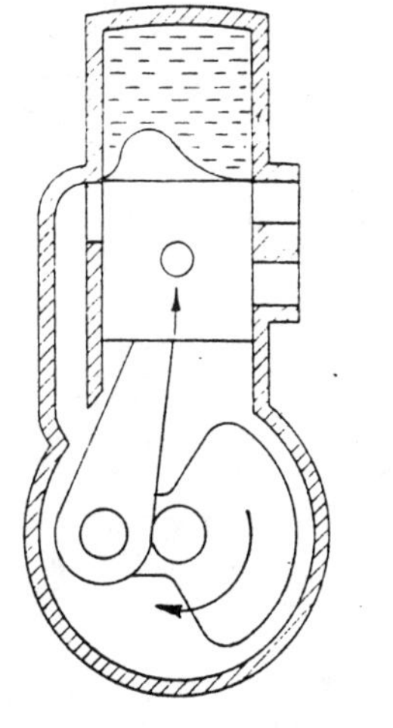

PISTON RISES COMPRESSES GAS AND RESTARTS CYCLE OF OPERATIONS

N.B. THE DEFLECTOR HUMP ON TOP OF THE PISTON IS NOT USUALLY FITTED TO MODEL DIESELS

respect, and has the power to fly the weight of the ignition gear without being troubled by it. In the case of the speedboat, the lesser weight allows planing to take place more easily with a smaller hull.

Admittedly the diesel must weigh a little more than a similar petrol engine of the same c.c. in order to stand those stresses we have mentioned, but it is not as much as the lost ignition gear. This loss of ignition gear weight is most important in the case of midget engines, and it allows us to produce perfectly practical and reliable diesel engines of 1 oz. all up, including fuel ready to operate! A reliable and practical electrical gear at the very best for a similar engine weighs many ounces over and above the engine's weight.

For instance, the engine may weigh 1½ oz. The coil 1½ to 2 oz. (as much as the whole engine). The flight accumulator 2 oz. at the lowest. The condenser ¼ oz. and wiring ½ oz. These ignition weights are the same for a large 10-c.c. petrol engine or a midget of 1½ c.c.

The diesel also enables us to operate our models with far less equipment to be carried in the field, on the race-track or on the pond. There are no spare batteries or booster starting accumulators to cart about with one—a small bottle of diesel fuel is all that is required, together with a small spanner to change propellers should they break or require tightening on the shaft.

I need hardly stress that a model can be built much more quickly if there is no complicated wiring of ignition gear to carry out with its booster plugs and electrical time switch and so on. This simplicity also makes the diesel model a better beginners' model for boys, and it therefore widens the field of power-driven models. But it has its snags; particularly is this so in the larger diesels, which have to be swung over the high compression we have talked about in order to start. It is a simple matter for even a small boy to bump over the little diesels to start them. One further very great advantage of the diesel is its use for flying-boats, seaplanes and speedboats—there is no electrical gear, which usually gives endless trouble when faced with damp and water.

THE EXTRA POWER OF A DIESEL

The diesel has more power per c.c. than the petrol engine. This is a well-known statement made by a number of people who write about diesels. The question that one does not see answered is, how much more power?

I recently carried out a little experiment with a model engine manufacturing friend in order to obtain a good practical comparative test. We rigged up a static test machine on a ball-bearing slide rail, mounted on a base secured to the bench. We used a slide rail mounted on ball bearings to eliminate friction as far as possible in order to secure readings that would be sufficiently accurate to draw worthwhile comparisons. To one end of the slide rail we attached one hook of a spring balance. The other hook of the spring balance was attached to the rail guide on the bench. At the opposite end of the slide rail we had an engine mounting fixed, so that a running engine could pull the slide rail along its guide and extend the spring balance. It was therefore possible to take readings of thrust in ounces.

One realises that static thrust does not show the full thrust that an engine is capable of, when the propeller is screwing its way through the air, because a static engine has the blades of the propeller largely stalled, but if the best possible propeller to suit each different engine is fitted, useful comparisons can be made between different types of engines.

We first fitted a well-known example of a 4.5 c.c. petrol engine that could be called a very sound example of power producer of that particular cubic capacity. This engine gave us two pounds of static thrust on our test apparatus. A 2 c.c. diesel, also a good example of its size and type, was next fitted. The diesel gave nearly $1\frac{3}{4}$ lb. of thrust. Both engines gave their best thrust with 11 in. diameter propellers.

This test showed (*a*) that the diesel of less than half the c.c. of the petrol engine would swing a similar size prop to the petrol engine. An important point to remember when fitting propellers to diesels, because their torque is greater at slightly lower r.p.m.—and not, as is frequently stated, at very high r.p.m. The same applies to the full-sized engine.

(*b*) That although the diesel was of less than half the c.c. of the petrol engine the thrust was not far off that of the petrol engine, thus giving a very good indication of the extra power produced by a diesel engine of any given c.c. in comparison with that of a petrol engine. The petrol engine makes more noise, which impresses many people.

The above facts, of course, are borne out in practical flying models. For instance, we may say that a 2 c.c. diesel will fly a 5 ft. 6 in. span model with the same vigour as a 3.5 c.c. to 4.5 c.c. petrol engine. This is even more marked in the case of the very minute engines. A 0.7 c.c. diesel will, for instance, fly a model that at least a 1.75 c.c. petrol engine is required to fly. The diesel gains in the smaller weight of power unit that the model has to fly, because of the lack of electrical ignition gear. Exactly the same applies to boats and cars of the small type. The efficiency of the diesel is bound up in a suitable propeller or gear ratio, as we shall see later in this book.

Further useful comparison is gained by tests I made using a hard-chined planing speedboat hull which I originally designed for 4.5 c.c. petrol engines. After a few minor alterations to the propeller I found that a 2 c.c. diesel would plane this hull almost as fast as the 4.5 c.c. petrol engine, of over double its capacity.

A large elliptical-winged model aeroplane that I have will fly well, powered by 4.5 c.c. petrol engines of various makes, but will not normally rise off the ground unassisted by push. When fitted with a 3.5 c.c. diesel with suitable propeller, this model takes off easily without assistance. The subsequent climb in the air is more rapid than with the 4.5 c.c. petrol engines.

Practical comparative tests like these give a far more accurate picture than records of r.p.m. and h.p. so often quoted. The proof of the pudding is undoubtedly always in the eating. Mr. Curwen's diesel car described in the final chapter bears out these tests, in another field.

Some people claim that the diesel is immensely more powerful than the petrol engine of similar capacity, others openly doubt if it is any more powerful. Neither of these extreme views is correct. Those who decry the power of the diesel have either not tested a good example, or have tested it with an incorrect propeller, poor transmission, inefficient model, or they habitually run the engine on too high a compression.

When reading of static thrust results, and similar performance data, including design details, it is as well to remember that manufacturers and amateur constructors are always making progress, which has already improved upon the results mentioned.

The diesel has a characteristic graph curve ; that is, the steep rise to maximum output, followed by an equally steep drop. The curves of the larger sizes of engines are generally steeper than those obtained from the small diesels of 1 c.c. to 2 c.c. A marked point is that the maximum B.H.P. usually lies at a lower reading with the larger diesels.

PROPELLERS FOR DIESELS

It is very important to realise that the diesel likes, and in fact requires, a larger diameter and blade area propeller than the petrol motor. It is a good point, I find in practice, to keep the diesel propeller of large diameter with a *low pitch*.

Before I leave our static thrust test apparatus, it is of interest to record that my friend and I put on six 1½-oz. diesel motors of only 0.7 c.c. All these six motors produced a static thrust very near each other in the neighbourhood of 8½ oz. Some produced this thrust at higher r.p.m. with a smaller propeller of slightly greater pitch, the diameter being 7 in., whilst others used 8 in. diameter propellers of larger blade area and a finer pitch and lower r.p.m. We favoured the larger propeller at lower r.p.m. because it gave easier starting due to its flywheel effect during the operation of swinging the propeller.

A very important point that should always be remembered in connection with model diesels is that due to the higher compression to be overcome, a propeller of heavy wood or plastic material is most desirable. Laminated wood is perhaps one of the best wooden types. The new plastic " unbreakable " props are excellent.

This feature helps the starting and running of a diesel, because the weight gives a good flywheel effect to overcome the high compression. It is also, of course, most important to fit a properly balanced propeller. See Chapter V for suitable propeller sizes and shapes. Many people actually do run their diesels with lightweight petrol propellers. The fact remains that starting and running is improved if a heavier propeller is used on aircraft and a heavy flywheel in boats or cars. One well-known modeller drills holes in his diesel propeller blades and fills these holes with little rounds of lead !

THE DIESEL'S PISTON

A normal two-stroke petrol engine usually has a deflector hump on the top of the piston to deflect the flow of gas from the transfer port up to the cylinder-head, whilst the exhaust gases escape through the exhaust port (see Fig. 2); on the other hand, most model diesel engines fit a plain flat-top piston which compresses or squeezes the gas up against the cylinder head or contra-piston. A flat top piston is fitted because of the difficulty of making a humped piston with the very small space available in order to obtain the high compression ratio required. This difficulty has been overcome in the "Morin" French engine in a rather novel manner. In this case a humped deflector piston head is used and it fits into a specially-shaped head (see Fig. 11).

The flat-topped piston works quite satisfactorily on model diesels, and is normal practice. Fig. 5 shows the flat-topped piston of a normal engine. The transfer port should not directly face the exhaust port.

The contra-piston and its function is explained a little farther on under the heading "The Contra-Piston," see also Fig. 3.

Pistons on small model diesel engines are normally plain and without any piston rings. They have to be an absolutely perfect fit with no appreciable tolerances. Many people talk gaily about the tolerances between piston and cylinder walls—the fact is they do not exist for practical purposes if the engine is to be of any real use !

It has been found recently that diesel engine pistons without any oil grooves cut in them are usually superior to those with grooves.

A rather staggering fact that may interest the slide-rule enthusiasts is that due to the short stroke of a midget motor of about 1 c.c. the piston travels only at the very slow *average* speed of 6½ m.p.h. at 7,000 r.p.m. The wear of these miniatures should therefore not be as great as a full-sized car engine with its very much higher piston speed.

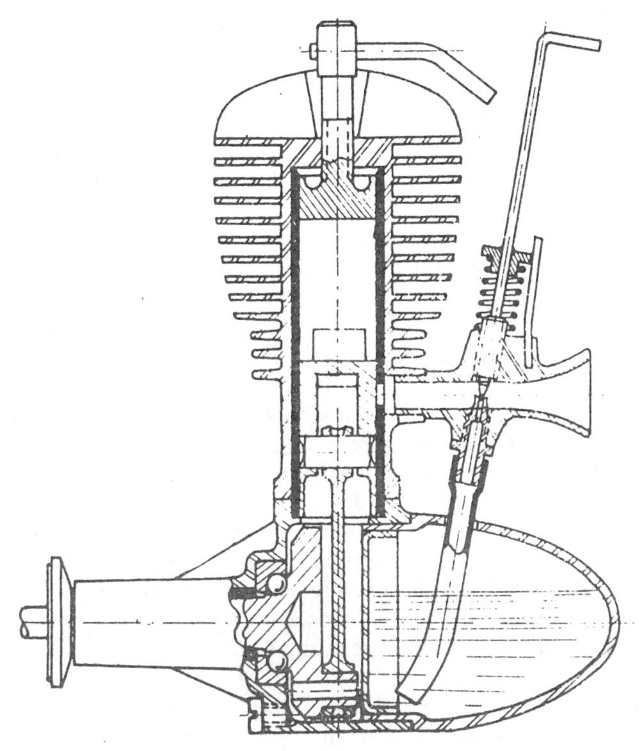

Fig. 3. A typical two-stroke model diesel engine fitted with a contra-piston to alter compression. The contra-piston can be seen at the top of the cylinder.

THE TWO-STROKE CYCLE OF OPERATIONS

Before we proceed further with the design features of the diesel it is as well to understand clearly the cycle of operation of the two-stroke engine. See Fig. 1 and Fig. 2. In spite of its simplicity and the fact that it has only three moving parts, the two-stroke and its method of working are not too easy to grasp.

The main essentials of a single cylinder engine are a piston that slides up and down inside a cylinder, a crank, and a connecting rod, which connects the piston with the crank. We have only to imagine the piston pushed downwards to realise that when an explosion takes place in the cylinder the crank automatically revolves just like the pedal and crank of a bicycle (see Fig. 1).

It will be seen that every time the crank makes a complete revolution the piston travels once down the cylinder and once up the cylinder. The piston therefore completes two strokes for every revolution of the crank, and there is an explosion every second stroke.

This is why we call the engine a " two-stroke."

The engine is kept going during the strokes when there is no explosion, by the momentum of a flywheel or a propeller. For aeroplane work we do not use a flywheel and that is why we must use a heavy propeller, which, in conjunction with its air resistance, acts as a flywheel. *In the case of the diesel this should be heavier than a petrol engine because we have to overcome a larger compression than in a petrol engine.*

In a two-stroke engine we do not use poppet valves to allow the gas to enter the cylinder and to escape after the stroke. Instead, we use ports cut in the cylinder together with a transfer passage between the crankcase and the cylinder, because we use the crankcase to take in gas and transfer it to the cylinder. By using the crankcase in this manner we are able to cut down the strokes from four to two per explosion.

THE ACTION OF A TWO-STROKE ENGINE IS AS FOLLOWS

This should be followed through step by step with Fig. 2. The sketch shows the normal petrol two-stroke engine as fitted with a " deflector " hump to the top of the piston, because this makes the flow of the gases easier to follow. Many model diesels are not fitted with this deflector-headed piston, as already explained, but the action is similar, although the ports do not face one another.

(1) Assuming that the engine is being rotated by swinging the propeller (or the flywheel in the case of a boat); the piston ascends in the cylinder and causes a powerful suction in the crankcase (which is hermetically sealed by good fitting bearings and joints).

(2) When the piston almost reaches the top of its stroke an inlet port in the cylinder lower wall is uncovered by the bottom edge of the piston. Atmospheric pressure then forces a full charge of mixture from the carburettor ("mixing valve") to the crankcase, because of the partial vacuum in the crankcase due to the suction mentioned above.

(3) The piston immediately closes this port upon commencing its descent, and the charge in the crankcase is then compressed.

(4) When the bottom of the stroke is almost reached and the crankcase compression is at its maximum, the top of the piston uncovers a transfer port in the cylinder wall (somewhat above the inlet port). This transfer port communicates with the crankcase from which the compressed charge is instantly transferred to the combustion chamber (i.e. the cylinder proper).

(5) The transfer port is closed again by the further ascent of the piston, and the charge compressed ready for the "explosion" to take place.

(6) *In a petrol two-stroke engine* the spark is timed to occur at approximately the top of the stroke, and the piston again descends by the force of the explosion.

In the model diesel engine the temperature of the mixture is raised to a sufficient height by compression to fire the charge at approximately the top of the stroke, and the piston again descends by the force of the explosion.

(7) Before the piston has fully descended, its top edge

uncovers a large exhaust port in the cylinder wall and the burnt-up gases escape by reason of their own force reducing the pressure in the combustion chamber to approximately that of the atmosphere, helped by the " extractor effect " of the exhaust snout or pipe if fitted.

(8) Immediately this condition has been reached the transfer port is again uncovered (it being almost at the bottom of the stroke) and as explained in paragraph 4 a fresh charge of explosive mixture fills the combustion chamber.

The cycle of operation then proceeds again as explained in paragraph 5 and onwards. It will be appreciated that when the engine is running the induction takes place simultaneously with the compression of a previous charge (paragraph 5). Similarly, the compression of a crankcase charge (paragraph 3) takes place simultaneously with the explosion of a previous charge (paragraph 6).

Thus there is one power stroke to every one revolution of the crank-shaft, or two strokes of the piston.

Two-stroke engines of the normal type of porting as shown in Fig. 2 and the " Majesco " engine, Fig. 5, can run in either direction—engines with ports such as the " Micron " can only be run in one direction ; usually counter-clockwise.

THE CONTRA-PISTON

Should he not have already done so, I hope that the reader will now understand how the two-stroke functions, and also that the model diesel works as a two-stroke with a very much higher compression than the petrol two-stroke, in order to fire the mixture without any electrical ignition equipment.

The next important point to be fully appreciated is how the compression ratio can be altered and why we should want to alter it.

If the fuel used never varies at all in its composition when mixed, then we can decide upon the exact compression ratio that is required to ignite it. There are engines on the market with fixed heads that cannot be altered. These are in the minority because of this very difficulty of always ensuring that the correct proportion of fuel constituents is available, and because people are not always accurate in measuring out fuel constituents ; also because when the engine warms up, the gas expands and alters the compression ratio.

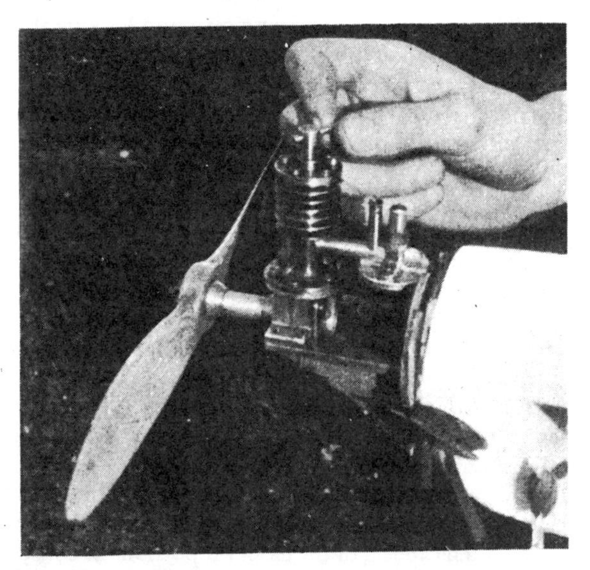

Fig. 4. Operating the contra-piston adjusting lever.

If a fixed-head diesel is owned, the operator must be very careful to adhere to the exact proportions and the type of fuel laid down. He cannot, of course, make minor adjustments to suit a different fuel or even loss of compression due to wear. Such an engine is naturally more simple to manufacture, but is limited in the fuel that it can burn, and is penalised by any carelessness in mixing up the fuel as laid down to be used. There is one further limitation in connection with an over-rich mixture that may be accidentally sucked in, but we will deal with the overcoming of this difficulty later on in the book.

The majority of engine designs are provided with an adjustable compression by the means of a lever or "tommy bar" fitted to the cylinder-head which screws an adjusting rod up and down through the cylinder-head. This adjusting rod either pushes down or allows the "contra-piston" to rise. When the tommy bar is screwed in a clockwise direction looking down on to the cylinder-head it pushes the contra-piston down. See Figs. 3 and 4, and Fig. 20.

The contra-piston in this case comes nearer the piston, and so raises the compression because there is a smaller space at T.D.C. for the gases to be compressed into. On the other hand, if the tommy bar is unscrewed in a counter-clockwise direction the contra-piston is allowed to be forced up towards the cylinder-head by the compression, thus allowing more space between piston and contra-piston and so reducing the compression.

It is absolutely vital that the reader should understand the working of the contra-piston and why it is used, so that later, when we come to the chapters on starting and operating the diesel, he will appreciate what is being said and what he is doing.

There are three main reasons why we may want to raise or lower the compression of a diesel—

(1) If we use a slightly different fuel mixture that requires a higher compression to ignite it and run it. People are also often careless of the exact proportions of constituents when mixing fuel. This is human nature and difficult to overcome. In this case, within limits, a slight alteration of the compression will usually overcome the difficulty.

(2) Many engines prefer a slightly raised compression ratio to start, with a slackening off of the compression when they become warm, the fuel then being more easily vaporised and the gas therefore more "expanded." On the other hand, I know of two engines, one foreign and one British, which prefer the operation in the reverse order, but this is unusual.

(3) Should too much fuel be sucked into the engine during starting operations, liquid fuel will often become compressed in the very small space between piston and contra-piston (frequently as small as 1/32 in.). This

liquid fuel is practically incompressible and the engine will become impossible to turn. *It will be damaged if it is forced over.* The contra-piston can be slackened back, and more space provided, the engine can then be turned until the liquid fuel is blown out of the cylinder. The compression can be returned to normal and a start made.

The reader will be reminded of the above three very important points when we come to the chapters on starting and operating the diesel.

A further form of compression adjustment is used in the case of the French "Ouragan" 3.36 c.c. diesel and the British "Airstar" diesel of 2.15 c.c. The cylinder-head is fixed, but the crankshaft is mounted in an eccentric crankshaft bearing. The shaft can, therefore, be moved towards or away from the cylinder-head by rotation of the eccentric bearing sleeve, which alters the space between piston and cylinder-head at T.D.C.

Fig. 5. Sectionalised "Majesco" diesel of 2-c.c. showing piston and contra-piston above it.

Fig. 6. French Micron 5-c.c. This engine does not incorporate a contra-piston.

The control lever appears similar to the variable ignition lever of a normal petrol engine. The advantage of this system lies in the reduction of the height of the engine. The " Airstar " which is in my stable works extremely well. Fig. 39.

THE VALUE OF THE CONTRA-PISTON

The operator can, with a contra-piston engine, be sure that he is obtaining maximum performance as his engine warms up by reducing the compression to its optimum value for warm running before releasing the model. It has been noticed on a number of occasions that fixed-head engines have faded out as they have warmed up and actually stopped before their fuel tanks have emptied. This, of course, is due to the rise of compression owing to the greater expansion of the gases as the engines warm up, which automatically raises the compression, which in turn acts as a brake to the engine.

Fig. 7. The famous baby 0.7 c.c. Mills has been reduced in size and price but increased in power. Note large exhaust port and clean robust design for only $1\frac{3}{4}$ oz. in weight.

COOL RUNNING OF DIESELS

The model diesel engine runs very cool; in fact far cooler than its petrol brother. I have seen several experimental diesels run perfectly without any cooling fins on the cylinder, and with only one fin on the head. Fins are eventually often fitted for the sake of conventional appearance because purchasers do not like curious and unusual things, a fact that has been proved in the full-sized motor-cycle, car and boat world. One can usually place the hand upon a diesel cylinder when the engine stops running. This cannot be done on a petrol engine cylinder unless the individual has a very hard and horny hand!

The Italian " Helium " diesel has vertical fins which make it look different, even though the performance may not be improved.

FRICTION IS THE BUGBEAR OF MODEL DIESELS

The elimination of friction as far as it is possible is most important to the diesel, even more important than on a model petrol engine, because any undue " stickiness " prevents the operator swinging rapidly over the high compression, and so

Fig. 8. The French 2.8 c.c. " Micron " sectionalised.

allowing the engine to carry on the swing with sufficient momentum for the first " explosion " to overcome the subsequent compression.

Fig. 9. The British Elfin fitted with a contra-piston and an adjustable compression ratio. Note the adjusting tommy bar protruding from the cylinder head. Weight $3\frac{1}{4}$ oz.

The problem is not an easy one for the manufacturer, because the piston and cylinder must be a perfect fit, and so must the mainshaft bearing in order to prevent leakage of the gas, *and yet the engine must be free to turn without any stickiness and still have good compression.*

It can be done, I know, because well-known manufacturers do it. If I were purchasing a new engine and spending " my all," I would not accept a diesel engine that was " sticky " to turn. I have had them, and they are the very devil to start. Neither would I accept an engine with poor compression.

HOW TO TELL A GOOD DIESEL WHEN MAKING A PURCHASE

When examining a new diesel, turn it round with propeller

fitted, see that it turns nicely and easily, but when it comes up against compression it must be difficult to bump over. Then try turning it against compression again, and hold it steadily with the compression half overcome. Now listen if there are any hissing noises of rapidly escaping compression, also observe if there is undue escape around the exhaust port from the piston; shown by a rapid bubbling of oil. The piston should hold compression for an appreciable time without the above signs of excessive escaping compression. Slight escape is permissible.

As I have already mentioned, the piston on most model diesels is of the plain ringless type and must be fitted "just so," with practically no clearance at all, and certainly not clearance that you can detect. Some of the latest American speedboat and racing petrol engines have piston rings fitted, to keep good compression when light alloy pistons are used in order to permit extremely high r.p.m. But the smaller diesels made abroad and in this country have plain pistons, because the r.p.m. of diesels is not so extreme. A good manufacturer knows how to fit a ringless piston to a cylinder bore, and he has the necessary machines to repeat the process in quantity. It is quite easy to make up one good engine by hand and think that the problem is solved. It is quite a different kettle of diesels when a number have to be made up for sale, and this is where many a budding new manufacturer failed in the early days of development.

Do not be misled by people telling you that a model engine should be really tight and difficult to turn when new, because it will free up when run in, and will therefore be a better wearing engine. This means that the manufacturer has shirked his business of obtaining a perfect fit and carefully lapping and running-in for you. The American engines of repute *all arrive free to turn, and yet with excellent compression.* Of course, I do not mean to say that a new engine will not be a little tighter than an old one, and that it will not benefit by careful running in; that would be incorrect. But the old idea of a new engine having to be stiff and "sticky" to turn is not now acceptable. *On the other hand, do not be caught by the engine that is easy to turn the whole way round the clock.* This means that the engine has not got

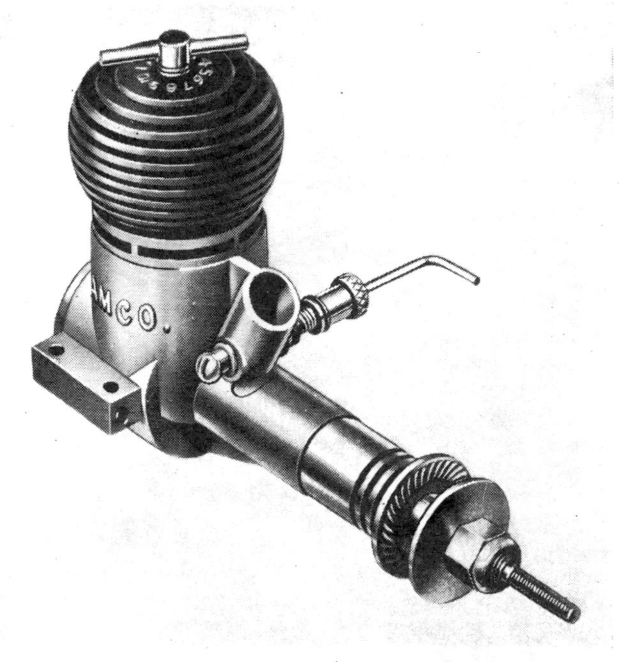

Fig. 10. The 3.5 c.c. Amco is a modern example of high efficiency, having a short stroke and big bore allied to large ports. This combination spells high r.p.m. Note ring of exhaust ports around the cylinder.

good gas seals and compression in the cylinder and crankcase. This engine will be very difficult, if not impossible to start, and it will be a poor runner if it does start.

One very soon learns to recognise a free-turning engine and yet one with good compression to be overcome. Provided that the purchaser knows that he has to look for these points, he will not accept the over-tight or the sloppy engine.

Many American and British manufacturers are now fitting ball and roller races to main bearings in order to eliminate friction as much as possible. This is a good point, provided

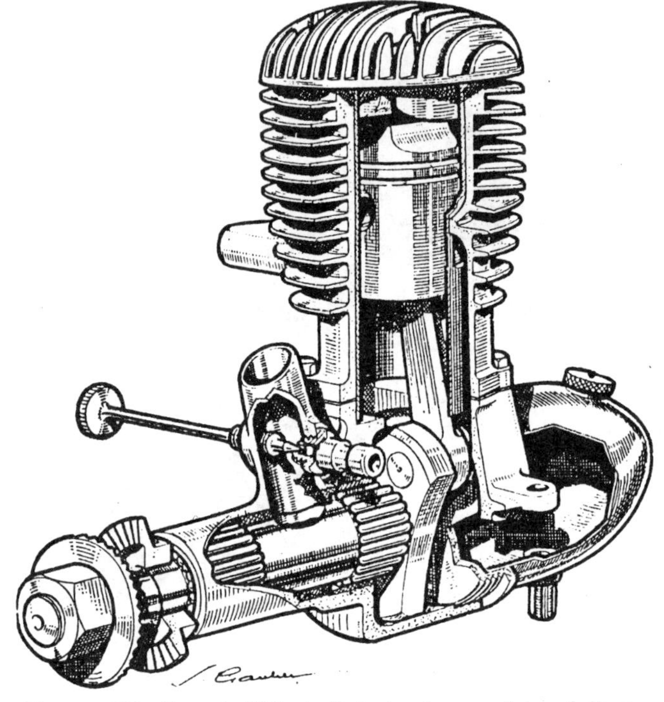

Fig. 11. The French "Morin" design has a piston deflector with recessed head. Note the roller main bearings.

there is provision made to keep good gas seals. This does not mean that a well-fitted plain main bearing is poor, for a carefully designed plain bearing gives every satisfaction for reasonable revolutions.

GAS-TIGHT SEALS

We therefore have the sliding up and down friction of the piston to keep down, yet we must keep a good gas-tight fit. We also have rotational friction to eliminate as far as possible, particularly in the main shaft, and yet keep a good gas seal to

Fig. 12. Monsieur Morin, of France, is responsible for this interesting engine fitted with a deflector hump and recess in the cylinder head.

prevent leakages of air into the crankcase, for you will remember that in Fig. 2 we saw how the crankcase on a two-stroke has first to receive the mixture and then compress it, and finally the descending piston pushes it up through the transfer port to the cylinder. If there are leaks of air to the crankcase the correct explosive mixture will be diluted. If there are leaks past the piston and cylinder wall the piston will not create a vacuum in the crankcase to suck in the mixture, nor will it pump it up from the crankcase to the cylinder *via* the transfer port.

It will readily be grasped that a two-stroke cannot abide air leaks, and apart from the piston and main bearing, there are also certain washers or gaskets and joints between cylinder and head, cylinder and contra-piston, cylinder and crankcase, crankcase door and sometimes detachable transfer port joints. *All these must be perfect and allow no leaks.* That is why I so heartily recommend the newcomer *not to take his engine to pieces or even allow an "engineering friend" to take it to pieces to have a look at the works.*

When he really knows what he is doing and how to make joints, it is a different matter. In the early days after purchase it is better to let the manufacturer have the engine back if it does not run properly. The French "Micron" engine instructions sum up this point very nicely—under the heading—

"*The life of your motor depends on this*:—The less you dismantle your motor, the longer it will maintain its quality."

RUNNING-IN THE DIESEL

How often one sees the wearisome spectacle of a new aeromodellist fiddling with a new engine in a new model on the flying field, whilst originally interested spectators become bored

Fig. 13. Running-in on a test stand.

Fig. 14. Testing the installation and running the engine on the fuselage, whilst the wings and tail are being made.

because of the abnormal time the poor fellow takes to start up his engine, and when it does start it constantly stops, and the laborious business of a restart has to be gone through again and again. All this is bad propaganda, and these discouraging results for the owner are due to :—

(*a*) not having a starting drill at his fingertips. This is given in Chapter IV—" On Starting."

(*b*) Lack of practice with the engine.

There is no advice so good as " know your engine." This can be done during the building period of the model aeroplane, boat, or car, firstly by mounting the engine on a test block of wood or a stand, and learning its tricks whilst running it in during the preliminary stage of building the model, and later by running the engine in when mounted in the completed fuselage, hull or car chassis, whilst the constructor is building the remainder of the model.

When running in an engine on the bench for a boat or car, an airscrew should be fitted for the purposes of cooling and load. Under no circumstances should an engine be gripped direct between the jaws of a vice. The crankcase is not designed to take this crushing strain.

It is far better to learn your engine in these two definite stages, because starting an engine in a fuselage near the ground is a very different thing from starting one comfortably swinging away high up on a bench! It also brings out any faults found in the mounting and general get-at-ability of the installation and controls, before the proud owner makes his, probably somewhat harassed, first flight under power, or his first voyage across the pond.

Figs. 13 and 14 show these two stages of running in, and will perhaps remind the new diesel modeller to carry out his running in and knowing his engine in two stages.

MOUNTING THE DIESEL

Owing to the high compression of a diesel, as a rule it runs more roughly than its petrol brother, and it will, therefore, shake itself loose unless well mounted. Swinging over high compression is also a severe strain on the mounting. It is therefore important to mount the engine rigidly and well with adequate bolts that will not shake loose.

Flimsy mountings cause vibration which upsets the mixture due to air bubbles in the fuel. Uneven running then ensues, and rapidly knocks bearings to pieces.

With regard to aeroplanes, although perhaps rather an ugly method, I know of no more practical way than the one I adopt. The mount for an aeroplane can be seen in Fig. 19. This is in the form of an elektron casting *that any engine can be firmly bolted to*. Elektron is a very light metal about 40 per cent. lighter than aluminium. The mount has a raised square cast on its rear. This fits into a square cut in the first three-ply former of the fuselage. The engine mount is then held on to the fuselage nose by elastic bands from wire hooks on the mount to wire hooks on the fuselage. These bands are made sufficiently taut to give in the event of a crash, and thus save engine and fuselage when the engine and mount are knocked off the fuselage, and yet prevent vibration of the power unit when it is running.

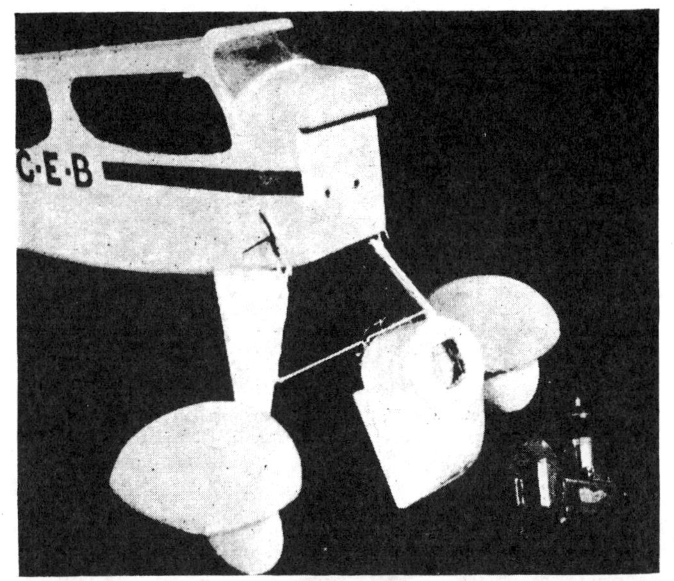

Fig. 15. The author's simple method of mounting a diesel. An elektron casting is screwed on to a solid laminated detachable balsa nose. This " distance piece " can be altered in length to suit various engines.

The mount, with its engine, can be detached quickly to change fuselages, or take to the bench. Down thrust or side thrust can be given or changed quickly when test flying a new model by simply inserting strips of packing wood between fuselage and mount. These can later be permanently secured by glue, silk and dope, when found to be correct.

I designed this mount many years ago and very many people have adopted it, or variations of the theme. The mount can easily be cowled in, to hide its ugliness, or left naked, if desired.

I frequently get letters asking me how a mount can be obtained. To forestall some of these—an elektron mount with a 3-in. circular backplate that will take almost any engine on the market today, can now be obtained from " BM " Models, 43, Westover Road, Bournemouth.

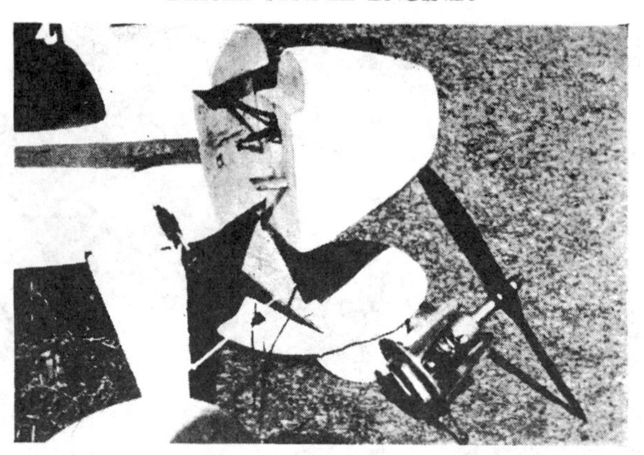

Fig. 16. Another view of the " distance piece " showing how it fits up to the rectangular nose of the fuselage.

Figs. 15 and 16 show another method that I often adopt whilst using one of these mounts. I screw an elektron mount on to a solid laminated and streamlined balsa nose. This in turn fits on to a square or rectangular fuselage nose. The balsa laminations can be made long or short to suit the weight of the engine fitted to obtain the correct c.g. position of the model. The whole is then held up to the fuselage by rubber bands. The mount knocks off in the event of a severe crash and will save damage to the engine.

If the reader elects to mount his diesel on wooden bearers in a rigid fixing to the fuselage in the American fashion, he should make sure that the bearers are square to the engine lugs and are really rigid, and do not give or distort the engine crankcase itself. The same applies to the wooden bearers in a boat or the bearers in a car. In a boat, oak bearers are generally used. See Fig. 17.

For those who like to use a simple wooden mount, but one that has the virtue of being detachable, I devised a scheme shown in Figs. 18 and 19.

The detachable engine mount is built up of plywood with a metal facing over the actual bearer arms. The two photographs will make the idea clear. It works very well in practice, although it is not quite so robust and firm as the elektron mount idea.

THE NEEDLE-VALVE CONTROLLED " CARBURETTOR "

The majority of model petrol and diesel engines have a very simple type of mixing valve in lieu of a carburettor. A full-sized carburettor is usually an instrument of chokes, tubes, jets and compensating devices to alter the mixture's strength under varying engine speeds and loads. The model diesel runs flat out, or nearly so. Its r.p.m. is nearly constant when it is flying

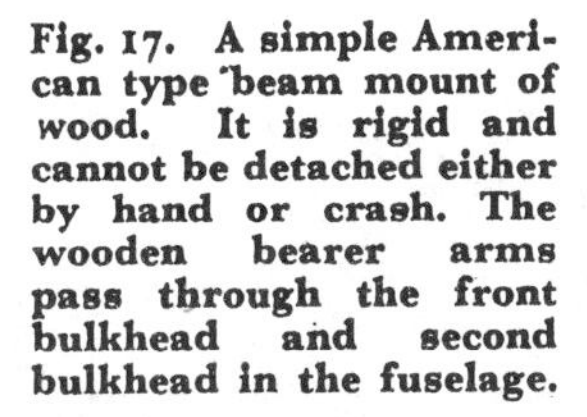

Fig. 17. A simple American type beam mount of wood. It is rigid and cannot be detached either by hand or crash. The wooden bearer arms pass through the front bulkhead and second bulkhead in the fuselage.

an aeroplane or propelling a boat or car. The question of compensating devices therefore does not enter into the problem except for radio-controlled models. A simple induction tube is therefore generally used with a jet projecting into the tube, the amount of fuel being regulated by a tapered needle-valve which seats in the jet orifice. When the thumb-screw of the needle-valve is turned clockwise, i.e., screwed down, the jet is progressively closed, and when the thumb-screw is unscrewed the tapered needle is raised from the jet seating and therefore allows more fuel to be sucked from the jet. See Fig. 22.

As already explained, the piston in the cylinder causes a vacuum or low pressure in the crankcase. This in turn sucks air in through the " mixing valve " intake tube to the crankcase. The air flowing at speed past the jet picks up fuel and atomises

Fig. 18. The author's detachable mount made from 3-ply, balsa fairing, and sheet metal, shown half detached. If the motor is inverted a hinged balsa top cowl can be added.

it on the way. An explosive mixture is produced which eventually is transferred to the cylinder and compressed. It then explodes or expands, and provides the power stroke.

The task of the mixing valve is to mix the correct quantity of air and fuel together to form a suitable explosive mixture which can be fired by the heat of the high diesel compression. In order to achieve this, the operator can alter the fuel to air ratio by the fuel needle-valve. (See Figs. 22 and 23, and refer back to Fig. 5, which shows a sectionalised carburettor.) It is possible to fit these miniature engines with a fixed jet, and so cut out all the bother of making a correct mixture each time. I have fitted up several engines with little carburettors with fixed jets and find it the best method for racing hydroplanes, but a considerable amount of patient fitting of, and experimenting with, various sized jets is required. Of course, once the correct jet is obtained, there is no further trouble over making the right mixtures. A fixed jet carburettor also must have a constant feed float chamber fitted, and the oil content in the fuel be rigidly adhered to.

All these complications require too many man-hours of production time for commercial engines built to a price, therefore the reader will almost surely buy an engine with the simple needle controlled mixing valve. This is satisfactory and therefore the instrument that we will discuss.

The secret of this type of carburettor is to find the correct opening of the fuel needle-valve, and then to mark the setting. The engine will always run at this, provided the fuel and oil content are kept the same. Only slight adjustment is required to correct the mixture to obtain maximum power, after the engine has been started, and when it is warm. If a different oil content is used this alters the amount of combustible fuel that can pass the needle and thus alters the fuel content of the mixture.

It should be remembered that dirt can clog or partially clog the needle-valve and seating. Fuel should therefore be filtered

Fig. 19. The author's elektron cast detachable mount can be fitted to a front bulkhead and cowled if desired. Compare this 5 c.c. Frog glow-plug engine's lesser height with diesels shown in this book of similar capacity, due to extra height of contra-piston of the diesel.

Fig. 20. The Mills 2.4 c.c. diesel sectionalised. Note the induction on this engine is by rotary disc valve at the rear of the crankcase. The transfer port can be seen at rear bottom of cylinder.

carefully and kept in a very clean container. If dirt is suspected in the needle-valve, unscrew it and blow through the jet and pipe, *but do not forget to return it to the correct setting for running*, i.e., the correct number of turns open.

FUEL FEED

It is poor practice to rig up a fuel tank with gravity feed, because the pressure of fuel, and therefore the flow of fuel, will vary as the tank empties. Fuel also dribbles into the engine when at rest and the crankcase may become flooded.

Fig. 21. Detachable mount of sheetduralumin.

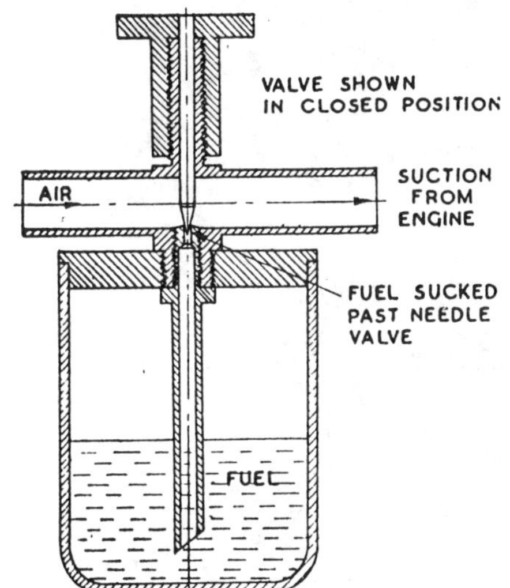

Fig. 22. Simple needle valve controlled carburettor.

The best method is to suck up the fuel from a small shallow tank located below or near the needle-valve.

The engine cannot flood itself when stationary, and because the fuel is near the needle-valve jet, the level of fuel and therefore the amount of fuel, will not vary appreciably when the model dives or banks, etc., in the air or gyrates upon the water. Because the tank is small, the level of fuel will not drop seriously.

See Fig. 25 to make clear the wrong and the right way of constructing and fixing the fuel tank. Control line flying requires a special tank to cope with the action of centrifugal force. This is fully dealt with in my book *Model Glow Plug Engines.*

Jim Walker, the father of control line flying and inventor of the famous " U Reely " control handle in which the lines are wound in and out as desired during or after flight, has recently produced a fuel regulator which is fitted between a special rubber tank shaped like a miniature hot water bottle and squeezed between two plates by a rubber band, thus giving pressure feed

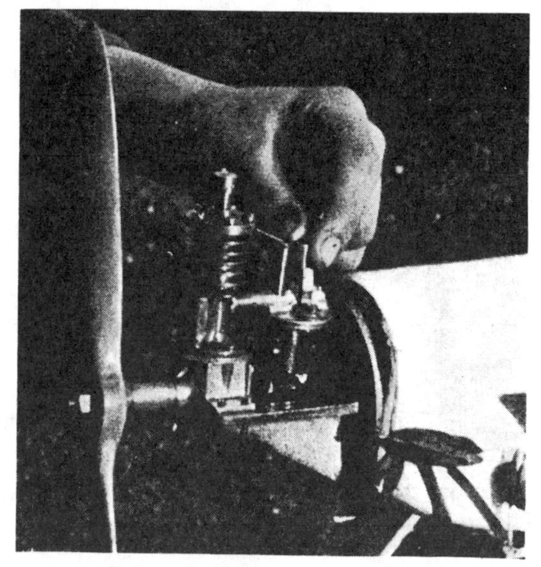

Fig. 23. Adjusting the fuel needle valve.

to the motor. The regulator meters the fuel by a diaphragm which insures absolutely regular fuel flow whatever the position of the model when stunting or radio controlled, and however great the action of centrifugal force on a circular path. This new feed is patented in America, and should eventually revolutionise model fuel feed reliability, thereby cutting out balky motors.

POINTS TO REMEMBER

Points that the novice should remember are that a suitable quick-burning explosive mixture of gas has to be made through the air being sucked past the jet, so that the right proportion of fuel and air is taken into the engine. He controls the amount of fuel by means of the needle-valve. He should keep to the best running setting once it is found, i.e. when the engine is two-stroking at its best performance. He should keep the

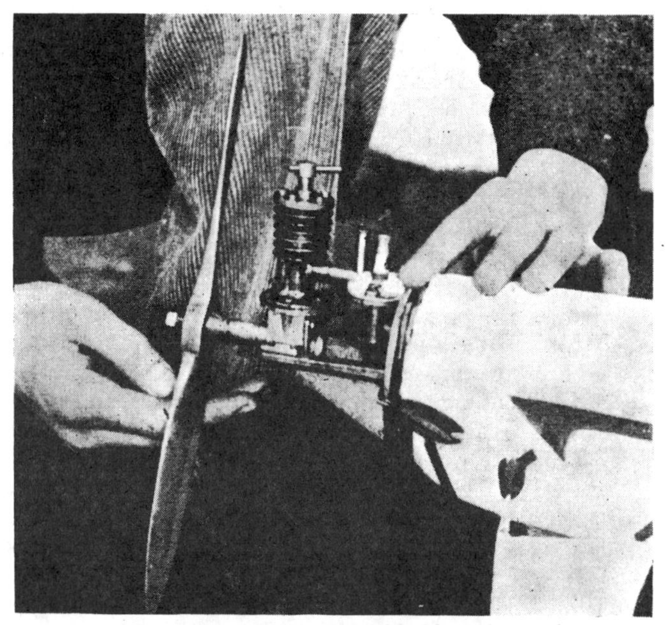

Fig. 24. When "sucking in" or choking the carburettor intake, the finger covers the induction pipe as shown.

needle-valve clean, and he can increase the amount of fuel taken in to start the engine by placing a finger over the induction pipe orifice and swinging the propeller once or twice. This must not be overdone or the engine will be flooded with liquid fuel (see Fig. 24).

SIZES OF MODELS CAN BE REDUCED THROUGH THE ADVENT OF THE MIDGET DIESEL

The diesel, with its elimination of electrical ignition gear, has recently brought the size of practical I.C. engines down with another bump. It is of interest to note that in 1914 Mr. Stanger's engine—which set up the first record free flight (petrol)—weighed around 2 lb. 12 oz. In 1932 I used a 2¾-lb. 28 c.c. engine to set up the first post-1914-1918 war record, and in 1933 I collaborated with Mr. Westbury, who produced a 1¼-lb. 14.2 c.c. engine, with which I set up the first really long record flight in history of 12 minutes 40 seconds "out of sight."

In 1934 the Americans produced the first real miniature engines of 9 c.c. and 6 c.c., the "Brown" and the "Baby Cyclone."

In 1935 the "Elf" people of Canada produced the first really baby commercial I.C. engine of only 2.4 c.c., weighing about 8 oz., with electrical gear and wiring.

The above were all petrol engines.

In recent years petrol aero-engines like the "Mighty Atom" and "Arden" from America and the British 1.7 c.c. "Frog" have been produced, weighing around 2 oz. bare, but had to add the weight of the coil, condenser, flight battery, and wiring, before flying could take place. These extras totalled up to considerably more than the engine weight.

Now, we have got down to diesels of 1 oz., and baby glow-plug engines with no other impedimenta, ready for flight. Fig. 26 shows my old 28 c.c. 2¾-lb. engine, that set up the first post-1914-1918 war record, side by side with a little commercially obtainable diesel weighing only 1½ oz., made by Mr. Colyer, which I have flown a considerable amount in different models, and also fitted an example to a tiny speedboat.

There are now a number of 2 oz. diesels obtainable. These are mentioned in Chapter II. Also see Fig. 7—The Mills 0.7 c.c. diesel. The new Albon Dart of 0.5 c.c. weighs only 1.2 oz.

The diesel has made the really midget internal combustion engine and model a practical possibility. Thus we have another milestone in the history of model I.C. power-driven craft.

The smallest known commercial engine, at the moment of writing, is the 0.16 c.c. French "Allouchery," which is alleged to run at 12,000 r.p.m., and can fly a model of under 4 oz. all up weight. This engine is too expensive to produce as a successful mass-produced commercial project. The "Allouchery" people produce a practical little engine of 0.7 c.c. This engine is also made with a specially lengthened crankshaft for those who like building scale model aeroplanes with a long streamlined nose.

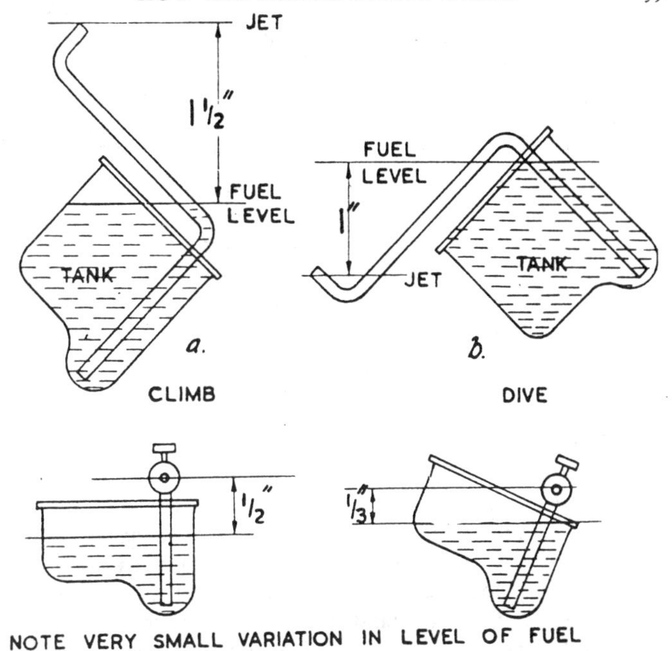

Fig. 25. The right and the wrong way of fitting suction fuel feed to model aeroplanes. Wrong method above, correct method below.

A phenomenally small diesel engine of only 0.04 c.c. was shown at the 1948 "Model Engineer" Exhibition by Mr. Fjellstrom, of Sweden. He demonstrated the engine in action. It ran so quietly that onlookers had difficulty in hearing its exhaust note above the noise of the exhibition!

THE LARGE-CAPACITY DIESEL

I have remarked upon the virtues of the very small diesel, and how the advent of the diesel has allowed us to make smaller model aeroplanes, boats and cars. I have also explained how, in my opinion, the petrol and glow plug engine scores in the larger sizes.

There is one aspect with regard to the larger diesel that I think we would do well not to overlook. The multi-cylinder diesel, in the form of a twin or even " three in line " engine, makes an attractive proposition, provided the compression of each cylinder can be accurately adjusted.

The chief advantage, apart from the added interest of the multi-cylinder, is that each cylinder will have the advantage of being in the small diesel class, with its ease of swinging over compression to start. There are already aero-twin model diesels in existence that work satisfactorily. Notably, a flat twin designed by the sponsor of the " Micron " and a vertical twin fitted to an Italian race-car.

THE COMPONENT PARTS OF A TWO-STROKE DIESEL

The component parts of a diesel are few and simple although they have to be very accurately made. They have to be more robust than a petrol engine of similar c.c. It is worth warning budding designers and constructors that they cannot usually fake a normal model petrol engine that has been correctly designed for the stresses of lower compression and convert it into a diesel. If a petrol engine will stand this treatment, it was probably designed unnecessarily robustly for its petrol mission in life.

DO DIESELS WEAR WELL?

I often hear this question asked. It is naturally a difficult one to answer because it entirely depends upon the design and the manufacturer, also how the engine is run after manufacture.

The diesels which I own that are what I call well-designed and made have improved with age, and as they naturally become the favourites, because of their sound starting and reliable running, they have had several years' use.

Provided a manufacturer builds his engine so that it is adequately stiffened up through the construction of crankcase and cylinder, and it has the correct metals in its design, with really adequate bearings, crankshaft and connecting-rod, there is absolutely no reason why the diesel should not last as long as its petrol brother. Connecting-rods should normally be made

Fig. 26. Dignity and Impudence! The "Wall" 28 c.c. beside the "Majesco Mite" 0.7 c.c.

from steel with hardened bearings. Some are made from duralumin, but, generally speaking, are better not cast.

If, on the other hand, the engine is not stiffly robust, and bearings and connecting-rod more suitable to the lower compression of a petrol engine are used, the life of a diesel will be short. The answer is quite simple. Buy a diesel made by a manufacturer with a reputation, and *then be careful not to run it with too high a compression setting.*

With regard to the main bearing, lapped cast-iron is often recommended in preference to the more usual bronze. It may, however, interest readers to know that I have an engine which has done a great deal of running with the crankshaft running direct in the aluminium alloy crankcase, and I believe that this will eventually become a normal practice in many model engines, because the softer and more porous bearing metal holds the oil and improves lubrication, apart from the very important advantage in reducing production costs. There are several firms now adopting this practice with perfect success.

WHY DOES A MODEL DIESEL LOOK TALLER THAN A PETROL ENGINE ?

The great majority of diesels are fitted with a contra-piston. This adds to the height.

Many diesels at present are of the long stroke variety in order to incorporate a long piston seal to obtain good compression and facilitate obtaining the high compression required. This long stroke naturally also demands a larger crankcase to allow for the longer crank throw. There has been a recent trend towards the short stroke " square " engine for high speed " hot " performance to suit control line models. A short stroke and bigger bore engine has many advantages. The well-known and popular " Frog 180 " diesel is a good example, also the " Elfin " and the " Amco." These engines run at very high r.p.m. for diesels. A newcomer is the Davies-Charlton 350.

One of the main advantages of the " square " big-bore short-stroke engine is the shorter distance that the piston has to travel. The piston speed is therefore not so great, which helps the mechanical wear factor and the attainment of high rotational speeds. The overall height of the engine is naturally not as great, which is an advantage for those individuals who favour engine cowling on their model aeroplanes.

The shorter stroke engine, being reduced in size, also means a reduction in weight. There is a lower inertia of the connecting-rod, and the gas flow into the cylinder is improved.

The main disadvantage of the " square engine," with bore and stroke approximately equal, is that the stroke being shorter, it makes the attainment of high compression more difficult than in the case of the long-stroke engine, because the combustion space is less compact. As a result, many model enthusiasts condemn this type of diesel. The secret seems to be a very carefully fitted piston, and vast exhaust ports. The 3.5 c.c. " Amco " seen in Fig. 10 forms a good example of a very high performance diesel. The long stroke diesel (such as the E.D.

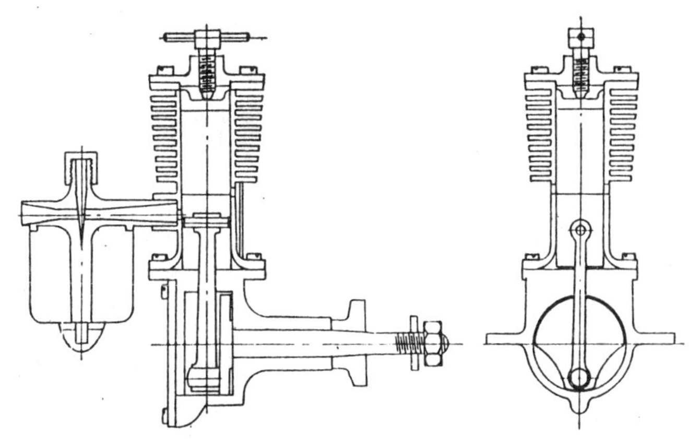

Fig. 27. The original 2-c.c. "Majesco."

Mark IV 3.5 c.c. engine) is a better "puller" as opposed to high speed "revver." (Fig. 50)

DIESEL DRAWINGS

Some readers with designing instincts will wish to study drawings of diesels. It is not possible in a book of this nature to reproduce the drawings full size. However, the reduced drawings of the simple 2.2 c.c. "Majesco" will give readers with a designing instinct some clues to get started with. (Fig. 27.) This engine has the simple cylinder induction port layout, whereas the engines seen in Figs. 10 and 11 are of the rotary induction crankshaft port type. The latter suits the high speed racing type of engine used for control line models of great speed.

SPINNERS AND FLYWHEELS

In the very early days of the diesel, it was considered necessary in many cases to fit a flywheel or a spinner with a groove in it to take a starting cord. This is now unnecessary and I would not own a diesel nowadays for aero work that could not be started by the simple swinging of the propeller.

The fact is that all the best British diesels start quite easily by swinging the propeller.

I remember one little British diesel that, when the first prototype was made, ran with most impressive power but would not start reasonably easily. I used to visit the manufacturer, and at one period, in those very early days of diesels, we thought that the only answer for these midget types was to fit a spinner with starting groove for a cord. This was being done by other people in England at that time on some larger diesels. We fitted this device and the engine started with reasonable regularity, but a cord is not really practicable on a model aeroplane. It is a nuisance, except in the hands of an expert, and inclined to become entangled with the propeller because it is so close to it and also tends to tear a very light fuselage to pieces. The value of fitting a baby diesel is lost if a heavy fuselage has to be made to resist the starting efforts!

We therefore decided that if one could not make a midget type start without such a handicap, the midget would be ruled out of court as a really practicable engine, and be relegated to the freak class. It was evident that the fuel did not become atomised and pumped up to the cylinder-head in that form from the crankcase. The whole problem was suddenly and very quickly solved by fitting a much smaller venturi tube to the carburettor, and hey presto, the engine started like a big-un, or very nearly so, and yet the power did not stop. I should mention that the engine was already an easy one to turn and therefore friction did not enter into the problem.

Now there we have a most valuable tip, for the man who fancies his chance at making diesel engines and who finds he must use a flywheel to start. I have experienced this in the case of other engines and am now quite satisfied that if any aero-engine *must* have a flywheel spinner to start by cord, it is an admission of failure in at least one direction.

The reader will probably have noticed that a number of the early diesels were fitted with these starting cord spinners. I can well remember in the early days of petrol engines, many years ago, having to fit starting pulleys. No one does this

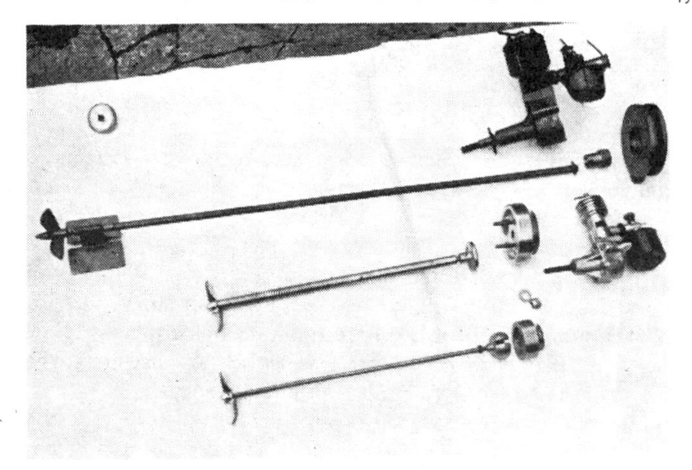

Fig. 28. A selection of boat transmissions. The stern tube is located through the hull floor with plastic wood. All flywheels have grooves for the starting belt. Universal joints are by nut and pin, or pegs and slotted disc, or by ball joint inside a recessed flywheel as in the "E.D. Bee" transmission seen at the bottom of the photograph.

nowadays. History has repeated itself apparently in the case of the diesel!

The cure is a combination of a good piston fit, easy turning of the engine, correct transfer port and correct venturi tube size of the carburettor, and, of course, sound port timing.

The flywheel I have developed for boat and car work for engines between 2.4 c.c. to 5 c.c. is made from brass. It weighs $\frac{3}{4}$ lb. and is $2\frac{1}{2}$ in. diameter and $\frac{1}{2}$ in. across the rim. It is not turned out in the centre other than to take the special coupling nut. (*See Fig.* 28.) A steel propeller shaft with a small steel ball and cross pin on the end fits into the coupling nut.

In Chapter IV I have included a photograph which shows how to start up a boat or car by cord. It is advisable to use a small driving peg fitted to the backplate on the engine shaft, otherwise a flywheel tends to come loose if a diesel backfires. On the other hand, I have found a large peg is too rigid for a wooden propeller, as it often causes it to fly to pieces, which can be dangerous.

MAKERS' RATINGS

I am never particularly impressed by advertisements or statements that the so and so engine will do 6,000 or 10,000 r.p.m. or is rated at 1/6 h.p. or 1/10 h.p. These statements mean little. Under what load did they do these "r.p.m." and how was the h.p. calculated? I prefer practical thrust comparisons or practical feats of flying model aircraft of certain wing spans and weights of propelling boats and cars of certain sizes and types at given speeds. Such data give a practical answer to what may be expected from an engine on purchase.

The most extravagant and meaningless claims of r.p.m. and h.p. are often made by makers, particularly overseas, and when I hear protagonists discussing these claims in supporting their favourite engines I am reminded of Mark Twain's remark, "Be honest. This will gratify some people and astonish everybody."

I am giving examples, in Chapter VI, of models that have been operated by different capacity engines.

"DIESELS RUN SLOWER"

It is most important for the user of diesel engines to realise that these engines develop their power at lower revolutions than the glow plug motor. The smaller the engine, however, the higher the r.p.m. of which it is capable, and the minute size diesel engines do actually revolve at speeds comparable with the average petrol engine. As their size goes up the maximum revolutions of which they are capable go down and it is quite easy to notice the difference in speed between a 0.7 c.c. engine and one of 2 c.c. Engines between 3.5 c.c. and 5 c.c. produce their power at considerably lower r.p.m. and swing a proportionately larger diameter propeller.

When 10 c.c. is reached, the r.p.m. is still further reduced and we are approaching the slow speed and high torque combination which is the characteristic of the full-sized diesel motor.

The very tiny engines have so small a combustion space that

combustion quickly spreads throughout the combustion area. The larger engines with their larger space take longer for the flame of combustion to spread. This feature is important, as it has a direct bearing on the propeller size which is employed (see Chapter V). A lower speed of revolutions and higher torque if properly applied through suitable propellers, etc., does not mean a lower power output. This has been explained earlier in the chapter under the heading—" The Extra Power of the Diesel." The recent development of the " square " short stroke big-bore diesels for racing models remarked upon in this chapter, brings this special type of diesel nearer the high r.p.m. of the petrol or glow plug engine.

THE THREE TYPES OF INDUCTION

The first is the normal cylinder port, as shown in Fig. 2, and also in Fig. 36, of the small Mills diesel in section.

The second is the rotary disc induction through the rear of the crankcase as seen in Fig. 20, showing the Mills 2.4 c.c., a long induction period can be arranged on this type. The " E.D. Bee " and the larger " Mark IV " also use this method. (See Figs. 47A and 50.)

The third method is the rotary crankshaft inlet port, much used by high revving motors. See Fig. 10, the " Amco " 3.5 c.c. diesel and also Fig. 9, the " Elfin " diesels which employ this type of induction.

CHAPTER II

BRITISH AND CONTINENTAL DIESEL ENGINES—PLANS AND CASTINGS

THE BIRTH OF THE DIESEL

The Swiss "Dyno I" was the first practical commercially-produced model diesel, which set the model world going on this type. It also set a fashion that is seen in a number of diesels made in different countries, namely, the long rectangular extension of the crankcase of Egyptian architectural appearance. As examples, the reader will see this feature in the Danish diesel "Mikro," the early British "Mills," the German "Eisfeldt" diesel, photographs of which are given in this chapter. There is something attractive in this shape but, of course, it is not vital to success, although it is a convenient method of stiffening up the engine. There are various other ways of attaining this object.

The war appears to have kept the glad tidings of the diesel's birth segregated to Europe, where development spread through France, Germany and Italy, and later to the Scandinavian countries.

The first serious British experiments seem to have occurred after the war, when the "iron curtain" of the Continent was raised. I was interested to learn that Mr. Lennan, of Scotland, first built a diesel in 1943, which was slowly developed over a period of two years. Mr. Sparey started experiments in 1944.

Mr. R. Trevithick is an amateur well known for his beautiful workmanship and novelty of outlook in designing model I.C. engines. Fig. 29 shows a very beautiful little engine made by him. The following are notes he has kindly sent me in connection with this engine, which shows amateur construction at its best.

"Engine, $\frac{1}{4}$ in. bore $\times$ $\frac{5}{16}$ in. stroke.

"Cylinder, alloy steel, machined from solid screwed into

crankcase 40 T.P.I.

" Transfer port and inlet brazed on.

" Crankcase, from solid dural, main bearing centrifugal cast-iron, burnished to size, as it is impossible to eliminate lapping compound from cast-iron.

" Crankshaft, from solid, 5/32 in. dia. in housing, tapered to ⅛ in. at front. Screwed 60 T.P.I. pin. case-hardened.

Fig. 29. Mr. Trevithick's miniature diesel, ¼-in. bore × $\frac{5}{16}$-in. stroke. An example of beautiful amateur construction.

" Pistons, first quality mild steel, case-hardened in electric furnace.

" Head adjustment, 3/32 in. dia. × 60 T.P.I.

" Connecting-rod, tool steel hardened.

" Carburettor intakes, venturi diameters, jet 0.003 in. hardened. Jet (adjustable), 3/32 in. × 60 T.P.I. hardened.

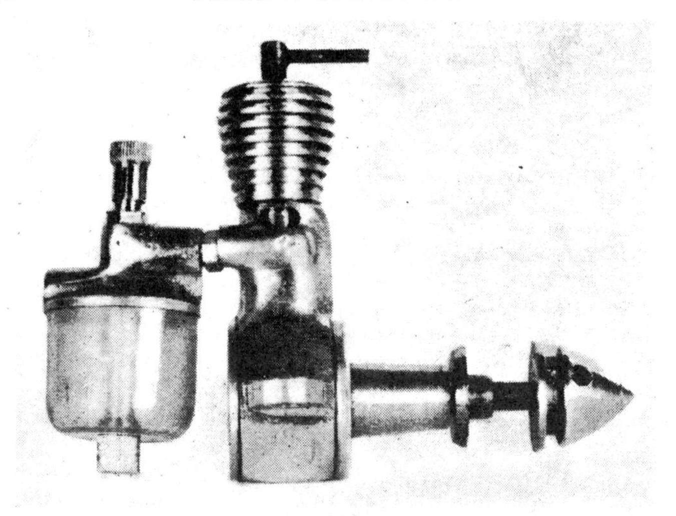

Fig. 30. A tiny commercial diesel with a reputation is the 0.3 c.c. Kalper, which drives a propeller of 7 in. diameter and 4 in. pitch.

"All lapping finished with magnesium oxide 0.0003 mm. grain, boxwood laps: rough lapping cast-iron or brass adjustable laps. All the work on this engine was carried out on a 3½-in. centre English lathe, now 16 years of age —except drilling the jet, which was done on a Wolf-Jahn watchmaker's lathe of 40 mm. centre."

Mr. Trevithick has some points to make which I will quote, because I feel sure the amateur constructor will consider them of interest, coming from such a well-known amateur authority, also because his remarks substantiate those I make farther on in this chapter in connection with amateur construction.

Mr. Trevithick says in his letter to me, "Last evening I got this latest effort running on 'Mills' fuel. At present it is touchy, but very fast—a little further work will have to be put in before it will 'pass,' I fear. There is a very tiny leak between barrel and the crankcase—apparently this should have been a lapped fit. The amount of lapping on these, and

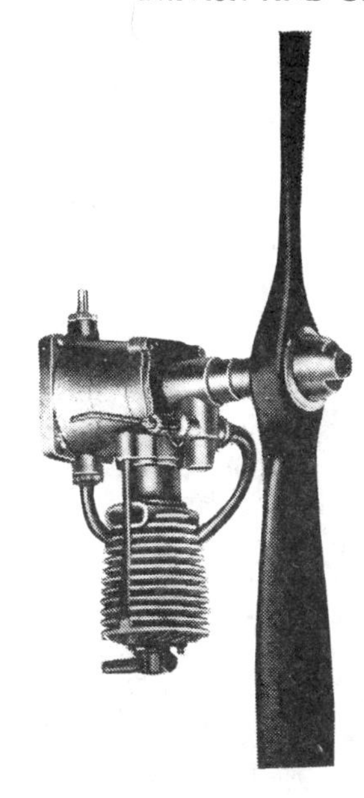

Fig. 31. The "Frog 100" is a 1 c.c. diesel fitted with a 9 in. diameter plastic propeller. The price is low and the performance is high. Weight bare 3.25 oz., static thrust 12 oz. +. A modified carburettor has recently much enhanced the performance. It is known as the "Spray-bar."

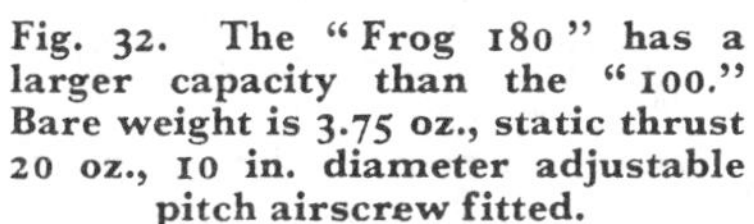

Fig. 32. The "Frog 180" has a larger capacity than the "100." Bare weight is 3.75 oz., static thrust 20 oz., 10 in. diameter adjustable pitch airscrew fitted.

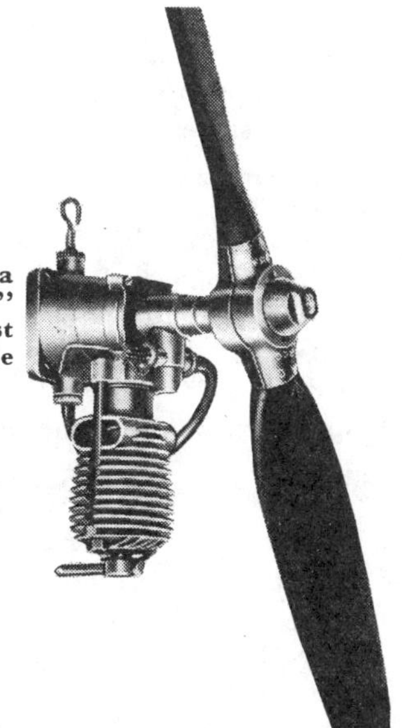

the time taken upon truing and making fresh laps is startling. The bore is ¼ in. and the stroke 11/32 in.; weight 1¼ oz. My lathe is now sixteen years old and I think a thorough overhaul of the alignments and headstock bearings seems to be called for, as the limits of these tiny engines are 'close.' I have made a very free-running rev. counter for these tiny engines—people are apt to be optimistic over engine revs. My experience suggests that engines with lower revs. are much easier to start."

BRITISH COMMERCIAL ENGINES

From the commercial angle, as far as I am aware, the first engines to be produced in this country for the general public were by the firms "Leesil," "Mills" and "Majesco." The latter two firms in considerable numbers. The "Leesil" took their engine off the market shortly after its inauguration.

Although we were late starting with the diesel in this country, we have already forged ahead. Our power output for weight of engine is, generally speaking, the best in the world.

The control-line craze in Britain has demanded very high power output for speedy models. This has acted as a spur to engine designers—as a result power output of model diesel engines in Britain has been raised to a very high pitch during the past two years.

The baby "Kalper" weighs a little over 1 oz., and has a capacity of only 0.3 c.c.—I have one of these delightful little motors. They have become very popular in the midget class. See Fig. 30. A 7 in. diam. by 2 in. pitch propeller is used for free flight. The Allbon Dart of 0.5 c.c. weighing only 1.2 oz. has superior power to the fractional c.c. midget American glow plug motors.

The British "Majesco Mite" of 0.7 c.c., weighing only 1½ oz., is an example of extreme lightness for its 8 oz. static thrust. Owing to its light weight it is a suitable power unit for models of the rubber "Wakefield" size—(Free flight). See Fig. 34.

The "Amco" 3.5 c.c. diesel is an ultra modern engine developed for very high speeds to suit control-line flying. I use my "Amco" also for free flight. It is one of the highest

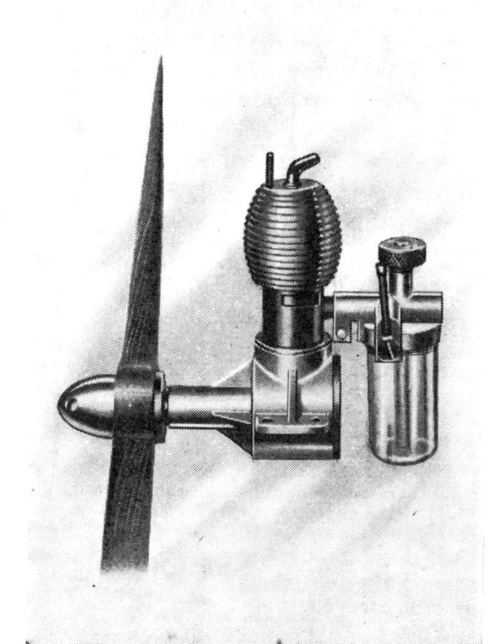

Fig. 33. The 0.87 c.c. Amco is a well known baby engine with a fine performance and $2\frac{3}{8}$ in. high.

speed diesels in existence. Exhaust porting is located completely around the cylinder with supporting bridges. There is an imposing inlet port via the crankshaft. On first viewing this diesel it is difficult to realise that such a comparatively large capacity as 3.5 c.c. in. can be housed in such a compact and small design. This motor must be allowed to do its work with a suitable propeller that permits high revolutions, for slower speeds kill output. The reason, apart from the porting mentioned above, for high power output with exceptionally high revolutions is that the stroke is short in relation to the bore. Thus with a short stroke the engine can rev. without excessive piston speed. The power curve reveals an output practically constant from 9,000 to 13,000 r.p.m., which is exceptional for a model

Fig. 34. The "Allbon Dart" of only 0.5 c.c. has exceptional power, capable of flying models up to 48 in. span. It is the British reply to the American Midget Glow-plug motors.

diesel and more in keeping with a glow-plug engine. A 9 in. by 8 in. propeller suits stunt control line work, whilst 10 in. by 6 in. is suitable for larger and lighter loaded C/L models. Speed models are suited by an 8 in. by 10 in. prop. The weight is 3.75 oz. without tank. Bore, .675 in. Stroke .5625 in., which makes a nearly "square engine." Refer back to Fig. 10 which shows this interesting engine.

The 0.87 c.c. Amco 87 has now firmly established itself as a particularly impressive little power producer and a good starter. A large number of these engines are already in circulation. Fig. 33 shows the clean design of normal 3-port type.

The "Frog" range of diesels are noted for their power, the introduction of plastic propellers, and their extraordinarily reasonable prices. I have used these diesels in many of my models, and watched their development from the early days of British diesels. They are first-class value for money.

The combined cone tank and engine mount makes them easy to bolt to a fuselage nose former. The tank permits upright or inverted mounting by simply changing the filler cap and the fuel line plug. A timer fuel cut-off can be provided integral with the filler cap. Control-line fans can mount their " Frog " motors, including the glow-plug " 160 " version, as " side-winders " with the cylinder on its side, and inverted flight can be indulged in, without resort to a special fuel tank, if the neoprene fuel line is taken one complete turn around the tank. This prevents surge of the fuel. For normal free flight, I personally mount my " Frogs " inverted. They are particularly good starters in this position as soon as one realises that flooding of the motor does not pay. I open the needle-valve one turn, suck in once only when fuel should be found on the choking finger, close the needle-valve, start up, and as the motor clears itself of fuel, open up to the normal run position of approximately half a turn open. A slight adjustment of the compression lever completes a sure operation.

The " Frog 100 " has a capacity of 1 c.c. and drives a 9 in. or 8 in. diameter plastic propeller. The bare weight of engine is 3.25 oz. The static thrust is well over 12 oz. This has been much enhanced recently by a modified and improved " carburettor " and fuel feed to needle valve, known as the " Spray-bar. " The rear of the cone tank has been thickened so that there is no difficulty in removing the filler cap. I fly my little 45½ in. span " Meteorite " and baby flying boat " Wee See Bee," described in the last chapter, powered by Frog engines. The firm have recently introduced an 8 in. high performance " unbreakable " propeller, also made from plastic material. Owners should always check balance of propellers as an unbalanced propeller causes bad vibration.

The " Frog 180 " is a larger and more powerful engine than the " 100," the capacity being 1.66 c.c. The motor swings a 10-in. diameter propeller, which in this case is adjustable for pitch by a simple clamping boss. This is a useful feature for changing to a higher pitch for control line speeds, which are usually higher. Appearance and dimensions are almost identical to the " 100."

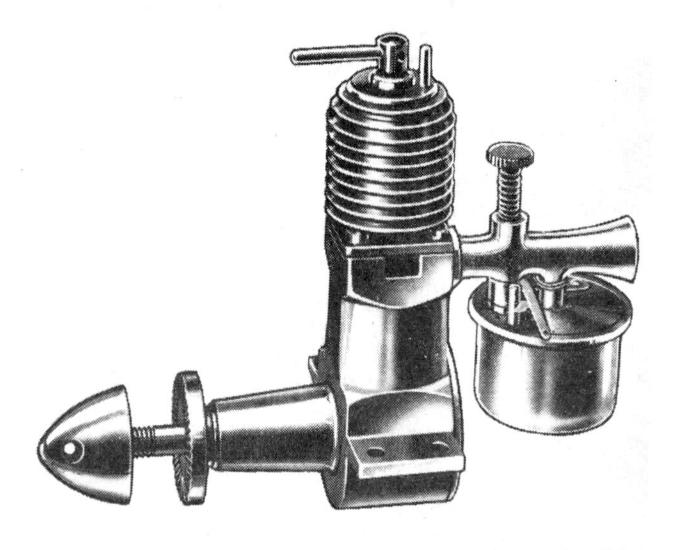

Fig. 35. The well-known " Mills " 1.3 c.c. diesel of British make has recently been much improved in appearance and performance, with a weight reduction to only 3½ oz. The static thrust is claimed to be up to 20 oz.

Figs. 31 and 32 show these engines. The adjustable pitch propeller will be noticed in Fig. 32. The " Frog 160 " glow-plug engine, which has an impressive performance is described in my book *Model Glow-Plug Engines*.

The new " Frog 250 " diesel is a particularly good looker with an excellent performance of the high speed variety. I particularly like the remote control needle-valve which saves possible damage to the fingers and makes for better fuel adjustment because the operator's mind is at rest. I also highly approve of the fitting of a fuel tank, which so many manufacturers evade under the umbrella that the owner will be only control line flying in all probability, and therefore can make his own tank. This is false reasoning, for even a control line fan likes to place his new motor on the bench and run it in with

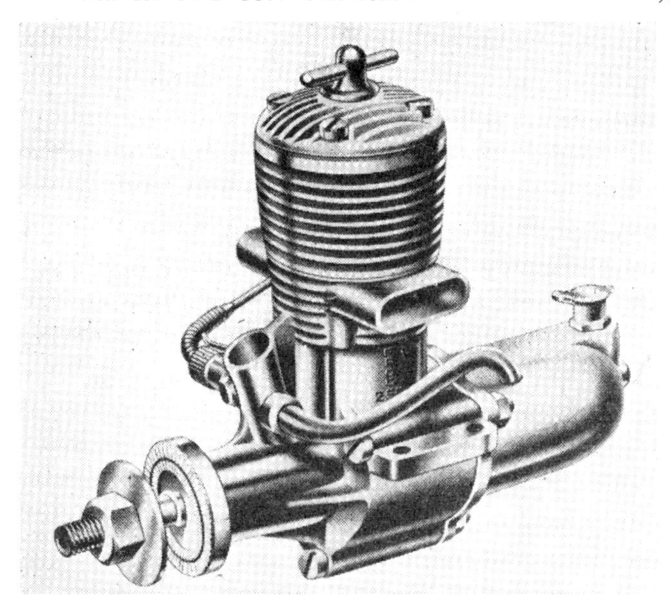

Fig. 35A. The new "Frog 250" diesel has a capacity of 2.49 c.c. Note the neat tank design and remote needle valve control.

tank conveniently *in situ*. The free flight enthusiast, and there are many still in existence together with radio fans, naturally wants a fuel tank ready built in. Furthermore, many a new modeller who buys his first engine has not the experience to build a tank or fit a commercial extra in the correct position for good flow of fuel. As a result bad fuel flow often gets an otherwise excellent engine a bad name. A tankless engine must reduce sales far more than some manufacturers imagine, and it is doubtful if the few pence saved is a sound policy.

Frogs do not fail in this respect, although their engines are sold at the most competitive prices, and what is more the tank is made so that it can be disconnected in a moment to connect up to a larger control line tank if desired. The fuel line is merely pulled out of the existing "250" tank, and the tank removed by a screw from the rear crankcase cover. The finish and appearance of this engine is good and on American lines. The bore is .580 in. Stroke .575 in. Cubic capacity, 2.49. Weight 5½ oz. Speed range, 2,000 to 10,000 r.p.m. See Fig. 35A.

The firm of Mills have a first class reputation from the early days of British diesels. The engines are noted for good finish and sound workmanship allied to easy starting. They are sold in three sizes. The baby, the middle size, and the larger 2.4 c.c. motor. Like every modeller, I have fitted many of my models with Mills engines. Last Christmas I was responsible for organising a five week demonstration of indoor round the pole flying and car running. I chose a 0.75 c.c. baby for the flying, and a 1.3 c.c. motor for the car. Those engines ran every day without a fault with absolutely reliable starting every time.

The famous 1.3 c.c. Mills seen in Fig. 35 started the firm's reputation. This engine has steadily been improved in appearance and power, over the years. Power exceeding 0.10 h.p. gives 9,000 r.p.m. with standard propeller. Weight is 3½ oz. Engines are claimed to be at their best after 400 hrs. running.

The baby 0.75 c.c. type makes an ideal little motor for small model aircraft and boats. The weight is only 1¾ oz. Bore .33 in. Stroke .52 in. Propeller 8 in. by 4 in. Max. power at 10,000 r.p.m. A new "popular" version of this motor is now produced without cut out at a cheaper price. See Fig. 7.

The larger Mills "Universal" 2.4 c.c. diesel departs from the firm's normal practice of induction system, adopting a rotary disc at the rear of the crankcase which can be seen in Fig. 20 and Fig. 43. This engine has an excellent performance with flexibility, making a good free flight engine for medium size aircraft and boats, as well as control line and cars. The bore is ½ in., stroke ¾ in. Speed 8,000 r.p.m. (10 in. by 5 in. prop.). Thrust 32 oz. Power 18 h.p. Max. power at 9,000 to 10,000 r.p.m. I have one of these engines fitted to my speed boat "Flying Fish" seen in the last chapter.

It would scarcely be expected that a diesel of the minute size

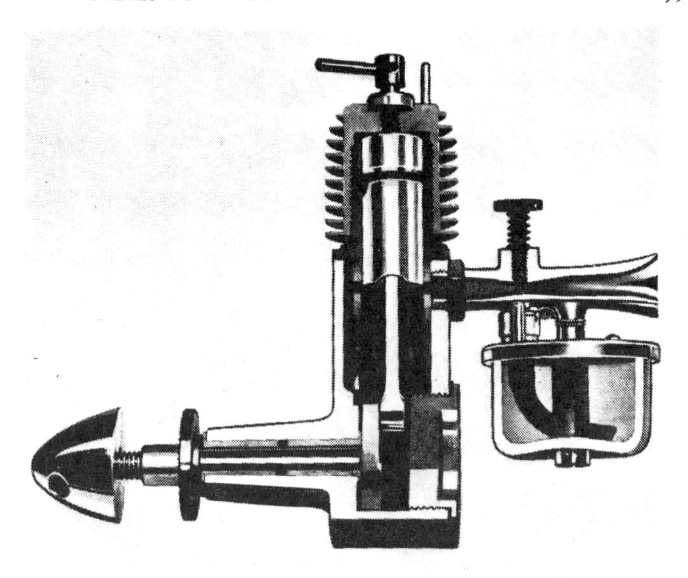

Fig. 36. This sectionalised view of the "Mills" diesel emphasises robust construction combined with light weight and clean design.

of only 0.2 c.c. could become a reliable commercial proposition at a reasonable price. And yet I have such a motor in my possession. It runs like an angry little insect to the end of each tank full of fuel. The makers suggest that a starting cord around the groove of the spinner should be used to start as the propeller weight, etc., is so small. After the first few days starting in this manner, which tends to wreck a very small and lightly-built model of the rubber duration class suitable for such an engine, I found that with reasonable care I could reliably start by swinging the propeller, which, by the way, is of only 5 in. diameter. It is a lovely little motor and is known as the "Hawk K" 0.2 c.c. diesel. The engine weighs 1 oz., having a bore of $\frac{1}{4}$ in. and stroke of $1\frac{1}{4}$ in.

The "E.R.E." (English Racing Engines) was specially designed for speedy control line work. This 2.48 c.c. diesel has

Fig. 37. Can you beat this? It is a practical commercially produced diesel of only 0.2 c.c. and drives a 5 in. diameter propeller with an exhaust note like an angry insect. The "'K' Hawk," Mark II.

an exhaust note not unlike a petrol motor in crackle. It has twin exhaust ports of very generous size and twin transfer ports. An air filter is fitted to the forward facing inlet. The carburettor can be changed to either side of the motor, thus making easy the mounting of the engine on its side, which has become a popular habit for control liners. There is a positive contra-piston stop lever which locks the contra-piston lever after the correct position has been found. The transfer of mixture from carburettor to combustion chamber is in an almost straight line. I have seen this little engine performing in a control line model with great zest.

The E.R.E. is also produced as a car unit complete with

back axle and two centrifugal clutches of great simplicity and effectiveness fitted into dummy rear wheel brake drums. The unit can be clamped straight on to a car chassis tray, and except for arranging a suitable fuel tank, that is all there is to it! Several of the Baigent cars shown in Chapter VI, are fitted with this unit which has been developed from the original B.M.P. by Mr. Baigent. Fig. 38 shows the car unit cum back axle. The clutch drums are attached to the back wheels, and therefore do not appear in the photograph.

A British engine having the feature of compression adjustment by eccentric crankshaft bearing is the "Airstar." This engine has a number of interesting features. No nuts or bolts are used in the assembly. Although the motor has a

Fig. 38. The "E.R.E." 2.48 c.c. diesel car unit comprises direct drive with two centrifugal clutches in the dummy rear wheel brake drums.

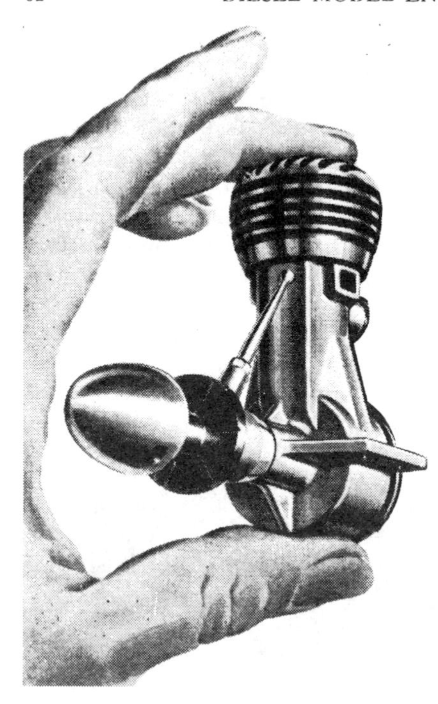

Fig. 39. The new "Airstar" of 2.147 c.c. has adjustable compression by means of the lever seen behind the propeller boss, which alters the crankshaft bearing mounted on an eccentric. The height of the motor is only 3¼ in., and it has a number of unusual features, including a fixed jet and adjustable air supply.

capacity of 2.147 c.c., the height is only 3¼ in., due to the eccentric bearing, which is made from gunmetal. The compression is altered by a lever behind the propeller, reminiscent of the variable spark lever of a petrol engine. This system eliminates the danger of breaking the connecting-rod or crankshaft by screwing a contra-piston too far down; furthermore, the contra-piston cannot leak or jam, because there is not one! Two tanks are provided. One suits ordinary work, and the alternative has gradations marked upon it to provide accurate measurements of fuel for competition work. There should be no flyaways using this method. A drain plug is fitted to the bottom of the crankcase. What a pity this is not a universal feature on all commercial engines. The German "Eisfeldt"

Fig. 40. The British 5-c.c. diesel "Eta" has contra-piston control, choke, and fuel cut-off, and a very fine "finish."

which I own is similarly fitted. Should one make a mistake and suspect an over-rich mixture, it is a moment's work to drain the crankcase and know how the mixture is when restarting operations commence. The "Airstar" has a fixed jet and an air throttle screw. The weight without its 10-in. diameter 6.5 in. pitch propeller is 5 oz. Altogether a most interesting engine. My "Airstar" is a good starter and a reliable performer. The unusual feature of a fixed fuel jet and controllable air supply works well in practice.

The E.T.A. "5" is a very well-finished British diesel of 4.9 c.c. Bore 0.6781 in., stroke 0.8593 in., weight 9½ oz., height 4.25 in. Consumption tests by the makers show an average running time of 1 min. per 3 c.c. of fuel. The tank holds 21 c.c. The induction has integral a sleeve choke fitted for remote control, and a fuel cut-off, for connection to a flight timer. Exhaust stacks and filler tubes are available in various lengths to suit cowling on a model. The contra-piston has a developed contour, giving a high efficiency combustion chamber.

Fig. 41. The "K" 5-c.c. "Vulture," Mk. II, has a high performance and was designed for radio control and speed C/L models, cars and boats. Note the easy mount flanges, large induction port and ring of exhaust ports around cylinder.

A static thrust of 32 oz. using a 13-in. diameter propeller of 7 in. pitch, is claimed for this engine.

The new "Eta" type R has a most attractive appearance with red cylinder head and a carburettor having the "new look." A belled mouth is provided, but the usual tank is lacking, and in lieu a suitable tube for the fuel line from a stunt tank is arranged. The owner can therefore fit whatever tank he favours for control line flying. This engine is one of the best-looking and largest capacity British diesels. I have found my "Eta" fly large free-flight models up to nearly 7 ft. span with unfailing

Fig. 42. This large 20-c.c. diesel was made by Mr. Court and is seen beside a " Frog " 1-c.c. The large engine is fitted to a full size dinghy hull of 10 ft. length, which it drives at approximately 3 knots. It answers the question—Can a model diesel be made successfully with a large capacity?

reliability. The same engine has now done a very large period of running, and starts as well, if not better, than when I had it new. The maximum brake horse power of the " Eta " has been discovered by an *Aeromodeller* test to be at 6,250 r.p.m. when a reading of 0.1805 b.h.p. was taken. The compression ratio is 12/1 to infinity. Airscrews : Free flight, 14 × 7 in.—12 × 8 in. Control-line, 12 × 10 in., 11 × 10 in.—10 × 12 in.

The " K " series of diesels include the tiny 0.2 c.c. already shown in Fig. 37.

The 5 c.c. " Vulture " seen in Fig. 41 is a high performance motor for speed models and also for radio control models, selling for a very reasonable price. The bore is ¾ in., stroke 11/16 in., height 3⅝ in., weight 7½ oz., r.p.m. with 10 in. by 6 in. prop. is 10,000 r.p.m. or 10 in. by 8 in. prop. 8,000. H.P.

at 9,000 r.p.m. is claimed to be $\frac{1}{4}$. Mounting by beam or radial.

The "K" Falcon 2 c.c. and the 1.9 c.c. "Kestrel" are practically identical in design details, having made their names control line flying.

Fig. 43. The disc valve "Mills" 2.4 c.c. engine is noted for its high performance and reliability. This engine can be seen in section in Fig. 20.

The "Falcon" has bore $\frac{9}{16}$ in., stroke $\frac{9}{16}$ in., height 3 in., with radial or beam mounting. I wish more designs would incorporate this choice of mounting, which is most useful and praiseworthy.

The "Kestrel" has bore $\frac{1}{2}$ in., stroke $\frac{9}{16}$ in., height 3 in., weight $3\frac{1}{2}$ oz. Both these motors do 9,000 r.p.m. on an 8 in. by 6 in. propeller or 8,000 r.p.m., using an 8 in. by 8 in. propeller.

The 2.24 c.c. M.S. is a lightweight of 2.8 oz. and claims to be the lightest of its kind. I have tried one of these engines and found it a good starter and reliable runner. The unusual tank around the crankshaft is neat and lends itself to easy mounting, particularly as the radial mounting lugs make it a simple matter to bolt the engine up to a firewall. If the reader will study the

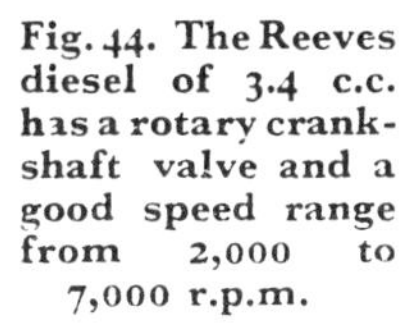

Fig. 44. The Reeves diesel of 3.4 c.c. has a rotary crankshaft valve and a good speed range from 2,000 to 7,000 r.p.m.

photograph of this engine he will note that the overhang is small, and the engine very compact.

The Foursome C.I. engine is made in limited quantities, but has a very good reputation and a clean appearance, as will be seen in the accompanying photograph. A contra-piston is fitted and the capacity is 1.2 c.c. The weight is 5 oz. and the claimed r.p.m. is 6,000 to 8,000, with an 11 in. diameter propeller. All bearing bushes, etc., are hardened, ground and lapped steel, as also is the cylinder and piston. I note with satisfaction that the makers wisely say in their instruction book, "Don't run your engine with contra-piston screwed down too far. You may get a few more revolutions, but a sweetly-running engine is impossible if you do."

The 1.2 c.c. "M.E.C." is a very light motor weighing only

Fig. 45. The new M.S. 1.24 c.c. diesel weighs only 2.8 oz. and has a neat tank located around the crankshaft. The mounting is radial and permits quick and easy bolting to a firewall.

1½ oz. The makers claim it weighs one-third less than the lightest of its approximate c.c., and can be used on sailplanes without any other alteration than for fixing. The engine has a high power weight ratio. It has even flown a lightly-loaded 7 ft. span model with success. Only M.E.C. fuel is supposed to be used with this motor. The bore is 0.450 in., stroke 0.460 in. Height 1⅞ in., length 3⅛ in., width ¾ in., diameter of bulkhead fixing flange 1¼ in.

The "Elfin" competition diesels of 1.8 c.c. and 2.4 c.c., have a very attractive layout and appearance rather reminiscent of the famous little Arden. The exhaust ports are in similar manner located completely around the cylinder. These engines have already gained a reputation for exceptional performance. The power of these motors can truly be called " terrific," and yet they are easy to start once one knows the technique.

The 2.4 c.c. " Elfin " is only a trifle heavier and larger in bulk than its brother of 1.8 c.c., but there is a big step up in power output. See Fig. 9. Both engines have done extremely well in control line events. I have used them much for high performance free flight and flying-boat work, as well as float plane models. One can not speak too highly of the " Elfin " engines. I now fly my little kit model monocoque high wing

Fig. 46. The Foursome is a nice looking engine with a sound reputation. The capacity is 1.2 c.c. and weight is 5 oz.

Fig. 47. The latest " Elfin " to suit Class I Competitions has a capacity of 1.49 c.c. and is fitted with the induction above the crankshaft. Engine weight is only 2½ oz. for very high power output.

Fig. 47A. The little 1 c.c. " E.D. Bee " is here seen in section. Note compact design, and the disc inlet port with induction pipe, needle valve and fuel tank located at the rear of the crankcase.

"Satellite" with an "Elfin" installed. The model can be seen in the last chapter. The latest range to the "Elfin" family is the new 1.49 c.c. motor for "Class I" competition work. See Fig. 47. This little motor has its induction arranged on top of the crankshaft, and naturally retains the highly successful feature of a ring of exhaust ports around the cylinder which is a family feature of these motors.

The capacity is 1.49 c.c., bore 0.503 in., stroke 0.460 in., weight 2½ oz., height 2¼ in., length 2½ in. A cast iron piston and main bearing are fitted.

The E.D. has become a most popular name amongst modellers, with a very successful sporting and competition reputation to its credit. The standard and first E.D. of the line is 2 c.c. and has a contra-piston adjustment made by means of a slot in the cylinder head. The head can be screwed up or down by a coin or screwdriver inserted in the slot. This feature looks neat, but in practice can prove rather a trial when the coin tries to jump out of the slot due to running vibration on a light model aircraft. The next engine is known as the "Competition Special." This is also 2 c.c., but has an enhanced performance. The compression adjustment is made in the more orthodox manner by means of a hand tommy bar. The ports are larger for increased r.p.m. Colonel Taplin's son set up an official control-line record for speed in the class using one of these engines. His speed was 89-95 m.p.h., which is no mean performance for only 2 c.c.! At first he could only get 55 m.p.h., but by careful trimming and other developments he gradually improved the speed capabilities of his model, until he arrived at the record speed, which was done in the still air of a hangar, so that a "flat" flight path could be maintained. The propeller had a pitch of 14 in., which naturally made the take-off rather slow. Speed had to be built up gradually. This model of the E.D. range also won the "Control Line Flight Competition" at the British Nationals, 1948. The static thrust is 23 oz. as opposed to the 18 oz. of the standard Mark II engine. I have used the competition E.D. a great deal fitted to a flying-boat of small dimensions. The model seen flying in Fig. 77A is powered by a "Competition Special."

The Mark III diesel is suitable as either a glow-plug motor or a diesel by merely changing the head. I have found that when using a propeller of 9½ in. diameter, 8 in. pitch, I get a very hot performance on my control line model. As a diesel I also power my control line flying-boat with this engine, which takes the model off water easily. The engine is capable of high r.p.m., and has special porting to that end. It created a record for class "C" cars of 41.7 m.p.h. in 1948 as a diesel. The capacity is 2.4 c.c.

An E.D. of only 1 c.c. has recently been developed which sells at a highly competitive price and has already become a best seller. It is known as the E.D. Mark I (BEE). It has a bore 0.437, stroke 0.400, weight 2¾ oz. This little engine has proved itself to be a very fine and reliable member of the baby class, with exceptionally easy-starting virtues.

The latest member of the growing E.D. family is known as the "Mark IV" diesel or "Three-Forty-Six." The capacity

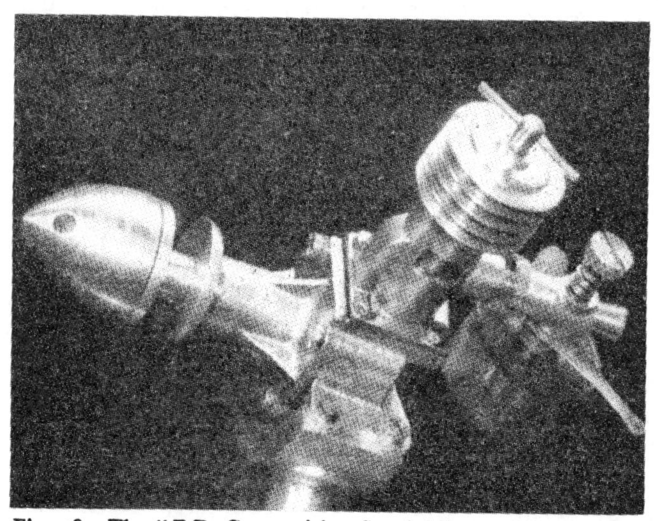

Fig. 48. The "E.D. Competition Special" 2 c.c. diesel has a line of competition successes. Larger porting is employed than on the standard 2 c.c. Mark II engine.

Fig. 49. The new E.D. Mark I weighs only 2¾ oz., and is 1 c.c. The engine has made a great name for itself owing to its robustness and high performance at a low initial cost.

is 3.46 c.c. See Fig. 50. This engine has a phenomenal amount of power for this medium capacity, and is a puller as well as a revver. A variety of propellers can be used to suit

Fig. 49A. The "Wildcat" is a Lancashire engine of practical layout, having a desirably long shaft to suit streamlining of the model's nose. An unmachined kit is also obtainable.

the job, which may be from control line to radio models, boats or aircraft. I have used one of these motors to fly a heavy radio model weighing 6½ lb. It makes an ideal medium size flying-boat motor.

A disc rotary inlet valve is used, arranged at the rear crank-case cover, and the exhaust porting is really generous allowing for very quick evacuation of gases. This is altogether one of the most useful diesel engines of the year. The bore is 0.656 in., stroke 0.625 in., weight 5¾ oz., height 3 in., width, 1 in., length, with extended propeller shaft, hub and spinner, 4⅞ in. The makers claim 10,000 r.p.m. and 0.250 b.h.p., which one would imagine is no idle boast. Propellers, free-flight, 10 in. by 5 in., stunt C/L 9½ in. by 6 in., speed 8¼ in. by 9 in.

The E.D. firm also manufactures one of the most reliable three-valve radio receivers, with transmitter giving off a constant "carrier" modulated by signal. I have used this set a great

Fig. 49. The Allbon Javelin, with a capacity of 1.49 cu. in. has proved itself a very "hot" little power unit of exceptionally low weight.

deal. It is the most simple for the novice to tune, and operates on the rising current system on receipt of signal to operate the relay, which can be made entirely foolproof if the operator will but trouble to ensure that his grid-bias battery is well up. See Chap. VI.

The D.C. 350-3.5 c.c. diesel is a newcomer on high speed, short stroke, big port lines. The induction is via rotary crankshaft valve. A fuel tank is located at the rear.

The "Reeves" 3.4 c.c. has a rotary crankshaft valve, and is suitable for planes from 2 ft. 6 in. span control-line to medium size free flight. The bore is 0.570 in., stroke 0.760 in., weight

Fig. 50. The E.D. Mark IV is a useful engine for radio control models and boats owing to its outstanding power for the medium capacity of 3.4 c.c.

Fig. 50a. The D.C. 350 diesel has a capacity of 3.5 c.c. It is very modern in design. The rotary induction valve, large ports and very short stroke all denote high performance at high r.p.m.

approx. 6½ oz. The cut out is of the positive valve type, and the tank is located at the rear of the crankcase. This motor has a good speed range. See Fig. 44.

The 5 c.c. " Wildcat " Mark III diesel has a greatly improved appearance to match its enhanced performance since inception. Design is orthodox employing well finished die castings. The two port exhaust system is used with direct entry inlet. This system usually produces reliable performance with good starting and power output at the medium speed ranges. This type of performance is particularly suitable for free flight, boats, and radio controlled models. On test the " Wildcat " was found to be an excellent example of its type and very well balanced. The useful speed range appeared to be between 4,000 and 10,000 r.p.m., using a 13 in. by 6 in. propeller. Weight with tank is 9.2 oz., bore, 0.687 in., stroke 0.875 in. Capacity 5.24 cu. in. See Fig. 49A.

The " Allbon Javelin " was developed from the very " hot " little glow plug engine, the " Allbon Arrow," an example of which I possess. The " Javelin " diesel comes into the competition Class 1, with a capacity of 1.49 cu. in. It is a pretty little engine, in appearance rather reminiscent of the " Amco " 3.5 c.c. motor, shown in Fig. 10. The weight is 2¼ oz., height 2⅜ in., bore 0.525 in., stroke 0.420 in. Free flight prop. is 9 in. by 4 in., C/L stunt, 8 in. by 5 in.

The " Javelin " is considerably more powerful than its glow plug counterpart, and develops 0.1 b.h.p. at 10,000 r.p.m.

CONTINENTAL AND AMERICAN DESIGN

The following diesels shown are a selection of typical continental practice which, together with those seen elsewhere in the book, will give the reader a fair idea of the trend of design overseas.

The first engine has an interesting history. It is the best known of the German designs, and was acquired after the war during the occupation of Germany from a Nazi youth leader who used the engine during the war for the inculcation of airmindedness in the German Youth under Hitler's regime. I now have this " Eisfeldt " diesel of 6 c.c. in my possession. It is a very powerful motor, and flies my large flying-boat of 7 ft. 6 in. span very nicely over Poole water. (See Chapter VI.) When in Germany, I discovered that the Mercedes car firm made a special model diesel engine for Goering.

The Swiss " Dyno I " is alleged to be the father of commercial diesels that led the way. Details of this engine have been copied by various designers, the most notable being the Danish " Mikro."

The French, like the Scandinavian countries, have exploited the diesel motor to the full, and have produced some very successful engines, also suitable working models for their diesels. One of the best known is the 5 c.c. " Micron, " a photograph of which appears at the beginning of this chapter. The " Micron " is also made in two other sizes, a 2.8 c.c. and a 0.8 c.c.—Fig. 55 shows the little 0.8 c.c. engine. " Allouchery " is another well-known French firm who are renowned for their excellent workmanship and sensible design. Various sizes of engines are made, and some have special long shafts to obtain a streamline effect in scale-type model aircraft with extended noses. Fig. 56 shows the 1½ c.c. " Allouchery." Monsieur Morin is responsible for a number of very interesting diesels of the fixed head and also adjustable compression type. The " Type 76 " is on the market. His other designs are available for amateur construction. The Italians have made diesels of varying design.

In America the sales and manufacture of model petrol and glow-plug engines have been prodigious and more than in any other country in the world. For some unexplained reason, the development of the diesel has lagged.

Fixed heads seemed to be popular at the beginning of American diesel effort, as the Drone and the " Mite " were two of the first to be advertised in American model journals. It is interesting to note that the 0.30 cu. in. displacement Drone weighs 11 oz. and has a compression ration of 18 to 1, whilst the " Mite " of 0.099 cu. in. weighs 2.6/10 oz., and has a compression ratio of 13½ to 1. Another of the first diesels to be produced commer-

Fig. 51. The author's German 6 c.c. "Eisfeldt" diesel.

Fig. 52. An early American representative, the "Drone," which weighs 11 oz., and has a displacement of 30 cu. in. with an 18 to 1 compression ratio and a fixed head. Compression ratio can be altered by fitting cylinder-head washers of varying thickness.

cially in America, but fitted with a contra-piston, is the " C.I.E." diesel which comes in the American baby class " A," for models around 48 in. wing span. See Fig. 52, which shows the " Drone." This engine has a really " hot " performance and the usual American outstanding finish. An alternative variable compression head is now supplied if desired. The new head gives first-class starting when using British fuels. The engine flies my 8 ft. span free-flight model, which is very good going for a 5 c.c. diesel. The Drone is the winner of many American " stunt " competitions (control-line), and won the 1948 radio-control contest of America. The latest Drone has a ball-bearing mounted crankshaft. Different thicknesses of detachable cylinder-head washers are used to vary compression ratio in order to suit varying fuels.

AMATEUR CONSTRUCTION

Certain readers will undoubtedly wish to make their own diesels. It is therefore advisable to examine what facilities there are on the British market at the time of writing this book that will be of help in this direction. A warning should be given that a high degree of engineering skill is required to build a model diesel, and there will be failures amongst those who try. A petrol model is quite a difficult proposition, and the diesel is even more so.

These words of warning will doubtless not deter the determined man. Indeed, I hope they will not do so, because even if there are failures a great deal of interest will result and a lot will be learnt. Those who succeed will be all the more satisfied with their achievement. I merely wish to warn the boy who has little spare cash and no experience, and few facilities, so that he will not be disappointed if he fails. In his case I believe it is cheaper to buy a made-up engine by a reputable manufacturer. Some of these are now sold at such very reasonable prices.

Mr. L. Sparey has produced a drawing and building instructions for a 5-c.c. diesel that can be obtained from the " Aeromodeller Plans Service." Mr. Sparey is a well-known engine designer, and was one of the first pioneers in this country to

Fig. 53. The Swiss "Dyno" alleged to be the father of all model diesels.

Fig. 54. The well-known French "Delmo" diesel, noted for its easy starting, 2.65 c.c.

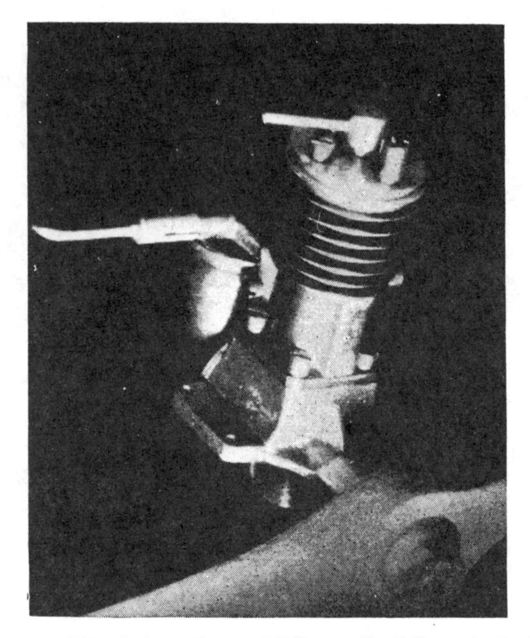

Fig. 55. The baby 0.8 c.c. " Micron " of French design.

build a diesel with which he gave a number of lectures and working demonstrations to various well-known model clubs and societies. As far as is known, he was the first one to design and make a model diesel in Britain early in 1944. He built this engine from " rumours " he got about continental diesels through soldier friends returning from abroad.

Recently, the " Masco " 0.8 c.c. diesel has been made available as a set of castings and working drawings. This little motor was designed by Mr. L. H. Sparey. The 2.8 c.c. Masco Buzzard diesel was designed by L. H. Sparey and D. A. Russell, and amateur constructors are offered diecastings comprising cylinder head and crankcase, etc.

A large number of Mr. Sparey's engines have already been

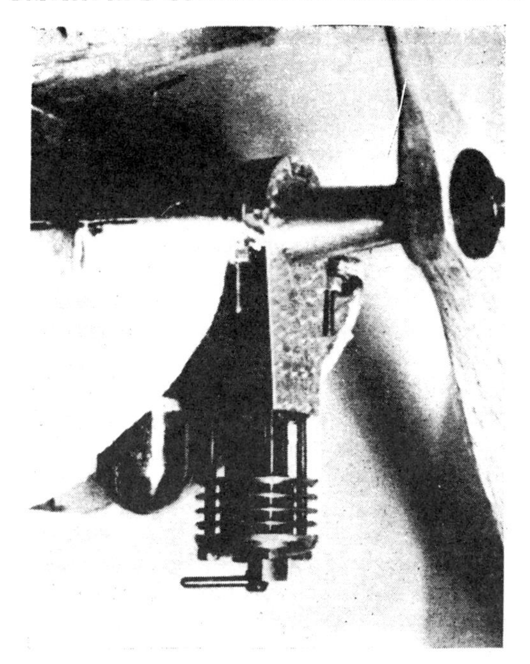

Fig. 56. The French 1½ c.c. " Allouchery," a very well-finished engine.

tackled by amateur constructors, and he is the first person to warn budding diesel builders of the pitfalls.

Although the parts are few and simple in a model two-stroke diesel, a high degree of accuracy is absolutely essential, particularly in the case of the cylinder, and the fit of the piston and the contra-piston. These latter components must be really perfect. Mr. Sparey's engine was made on a 3½ in. lathe and a drilling machine, therefore complicated machinery is not necessary. It is skill and care that is required.

Many amateur constructors will want to buy a set of castings, together with a plan. In this case, there is the 2.2 c.c. " Majesco " diesel which I have used so much in aeroplanes and boats. This firm are well known for their excellent and very complete set of 4.5 c.c. petrol engine castings which have sold in large numbers. The 2.2 c.c. diesel is kept fundamentally the same as the examples I have shown in photographs in this book, because it has proved so satisfactory. The only change is a cast cylinder-block, so that certain difficult operations for the amateur will be eliminated. Port timing, shape and size are exactly the same in order to retain the proved characteristics of the present engine.

There are now several other sets of castings on the market. Rapid progress of some manufacturers and the demise of others are two of the difficulties of writing books on engineering subjects.

Finally, there is really no reason why a thoroughly knowledgeable man should not have a cut at designing his own diesel and then making it. Keep it robust and remember the fit of piston to cylinder should be the theme to keep in mind. There are many reasons in favour of designing multi-cylinder diesels and also a four-stroke diesel where cost of production does not influence the choice of type. A four-stroke diesel with its positive exhaust stroke could well be silenced for boat work on ponds where the local residents object to noise.

NOTES ON THE CONSTRUCTION OF A 2-C.C. DIESEL

For those equipped with a reasonably accurate lathe of 3½ in. centre or more, the construction of a diesel motor is by no means beyond the capabilities of the model enthusiast. An amateur who has successfully made a petrol motor need have no qualms about attempting a diesel, bearing in mind that the higher compression ratio and violent detonation demands greater strength and soundly-constructed components. In addition, as the necessary compression of the fuel to firing point must be attained in the cylinder-head, it follows that little or no leakage down the piston can be allowed. Therefore the quality finish and fit of the cylinder bore and the piston must be excellent. The actual clearance between these two parts must also be as small as possible, and it can only lead to disappointment in the final performance of the motor if these points are not attended

Fig. 57. A neat Czechoslovakian diesel, the "Standard" 1.8 c.c. "Super Atom"

to. It is also necessary to use suitable metals for the cylinder liner and piston that will permit the above close clearances at working temperatures. For instance, cast-iron piston and steel liner.

Bearing in mind the above remarks, let us tabulate the main components that make up the diesel motor, with some suggestions as to the way to set about their construction. These remarks apply to the "Majesco" method of design, in connection with the Home Construction Set, which has been evolved especially for amateur craftsmen, employing as many diecast parts as possible. It is by no means the only way to construct a diesel, but will form a basis for discussion.

Crankcase

A diecasting which can be machined in two operations, drilling and tapping. The bearing should be reamed and lapped to ensure a good fit. Bronze or cast-iron bearing.

Crankshaft

This can be built up from stock material by screwing and silver-soldering. All joints should be absolutely clean and a low melting point solder used to avoid scaling.

Carburettor

A diecasting requiring very little machining. Three operations in the lathe, drilling and dieing the threads.

Cylinder

A diecasting with tubular liner insert. Three lathe operations, facing and boring to size. The liner is skimmed and shrunk into the cylinder casting, giving about 0.001 in. interference. Ports are cast in the cylinder, which should be arranged to align with those in the liner. Lapping the bore is left to last, and should be continued until a fine finish is obtained, together with roundness and parallelism. A series of brass plug gauges machined to a tight push-fit in the bore as lapping proceeds is the best way to test the condition of the bore. Thoroughly rinse the bore after lapping, and if possible blow out all traces of lapping compound. Be careful no compound remains in ports, otherwise it may become dislodged later and quickly ruin the engine.

Piston

Machined from the solid. Cast-iron. Remove the material inside and screw for mandrel. After drilling and reaming gudgeon-pin hole, screw on mandrel and turn to within 0.001 in. of finished size. (Obtained from size of the last plug gauge made for bore.) Lap piston with copper ring lap until it is a tightish slide-fit in the bore.

Contra-piston

Machine to a *light tap-fit* in the bore. Counterbore the top.

Connecting-rod

Machined from square steel to a tapered round section between big and little end. Drill and ream big and little end

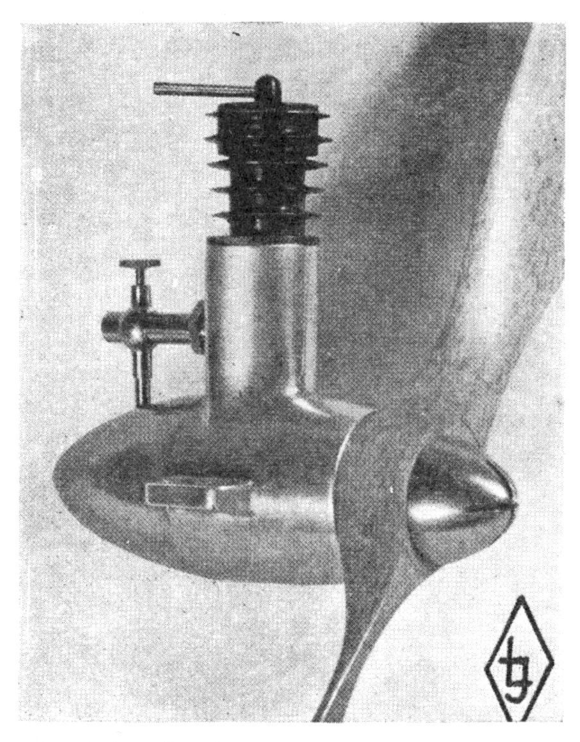

Fig. 58. A Scandinavian diesel, the "Monsun Standard," shows attractive design layout, and an eye for line.

about 0.001 in. small, finally lapping after hardening. This latter process can be carried out by heating to cherry-red and quenching, afterwards cleaning with emery cloth and tempering to very dark straw colour. If mild steel is used, this can be case-hardened only.

Cylinder-Head

Steel or aluminium threaded to take the contra-piston adjusting screw, and drilled clearance for holding down screws.

Propeller Backplate

Steel with taper hole to fit taper on crankshaft, and arranged to give clearance between crankcase bearing.

Crankcase Cover

Diecasting. This requires one lathe operation to machine spigot. Drilled clearance.

CHAPTER III

THE FUEL, LUBRICATION AND FUEL ACCESSORIES

GENERAL FUEL MATTERS

THE question of fuel for the model diesel is just as important as the subject of the compression ratio. In fact, it may be said that these two subjects go hand in hand, for one reacts upon the other.

It was the clever idea from the Continent of adding a suitable mixture of ether to the fuel that made the model diesel a possibility. Before this idea had been thought of, the model diesel in small sizes had been dismissed as impracticable owing to the weight and difficulty of fitting a fuel injector gear. Ether, as explained in Chapter I, increases the ignitability of the fuel, and permits auto-ignition without all the complication of injection of fuel at a predetermined point of the power stroke.

Having sacrificed the mechanically exact timed injection, it would be pardonable to expect that it would be impossible to obtain regular explosions when the piston arrived at the correct position at the top of the compression stroke. It might be thought that this would vary, but in actual practice it has been found to be completely consistent *provided the correct fuel, and a suitable compression ratio for that fuel, are provided.*

In fact, I would say that one of the outstanding features of the model diesel that particularly impresses itself upon my mind after many years of model petrol engine experience, is the extraordinarily even running of a well-adjusted diesel. Once started and running well, there is less tendency to change note and power output than in the case of the petrol engine, and the setting of the fuel needle-valve is definitely less critical than in the case of the petrol engine.

On the other hand, it is very important to correctly mix the proportions of constituents in the fuel, and to set the compression right.

The full-sized diesel is not fussy about what fuel is injected into its cylinders. It will stand almost anything. As most readers know, the normal diesel (full-sized) uses a crude oil that is refined far less than the spirit used in a petrol engine. The fuel is generally known as " diesel oil " or " derv." It is much cheaper than petrol, and also gives a better mileage per gallon when used in conjunction with a diesel engine.

We can use this " diesel oil fuel " in our models, but we must add a percentage of ether and a percentage of lubricating oil.

Paraffin and other fuels can also be used in lieu of diesel oil as the basis of the fuel, and there are special concoctions on the market sold by engine manufacturers.

My experience has been that the commercial mixtures which I have tried are highly satisfactory. On the other hand, almost all model diesels with adjustable compression will run reasonably well on a diesel oil fuel basis given later in this chapter.

I am considering the normal modeller's lack of facilities for obtaining ingredients, and also for measuring difficult percentages, when I sum up the situation in this chapter.

To cut a long story short, there are a considerable number of fuel mixtures that have been tried here and abroad. Many are most complicated, and the subject is shrouded in mystery by many people, with the result that there has been a lot of ballyhoo talked about these curious mixtures. Many, one suspects, have resulted from the difficulty of obtaining the simple ingredients that are obtainable in this country. I have tried most of these mixtures and found none better, and some not as good as the mixtures given in this chapter. This is not to say that we shall not improve our fuels for model diesels as time goes on, as there is a great deal to learn about the subject.

There are four well-known diesel fuels commercially blended, all of which I use at the time of writing. These fuels are blended by one or other of the large petrol companies, and each has an additive to promote easy starting and smooth running. They are known as (1) Mills Fuel; (2) Frog Diesel fuel; (3) E.D.; (4) Mercury diesel fuels. There is one Mercury fuel that requires no ether added. This is known as " Mercury No. 6," and is as easy to start as the normal etherised fuels. Normally, ether must be added to all diesel fuels. There are also a number of special blends made up by model shops which are quite satisfactory. Whenever I have examined damaged engines due to corrosion, this has usually been traced to the use of improper ether. The reader will do well to take to heart the advice given later in this chapter regarding ethers, *and under no circumstances whatsoever to use what is known as "commercial ether."* If unable to obtain any of the above fuels for any reason, I use my simple mixture as a second choice. It works very well but has no additive such as amyl-nitrate.

It is important, however, that, if the reader has a "fixed head" engine, he should use only the fuel laid down by the makers, because the *unadjustable compression ratio* has been designed to suit that fuel and no other. There are, however, few "fixed head" engines on the market.

One must admit that some of these "fixed head" mixtures have rather difficult percentages to measure out.

Let us examine these fuels so that the reader can use whichever he fancies, or can easily obtain. We will then end up by giving the "fixed head" fuel of the well-known French "Micron" engine, as an example for a "fixed head" engine, remembering that each "fixed head" engine has its own special mixture. One of the advantages of the variable compression engine, to my way of thinking, is that it is far more accommodating as regards the mixtures which may be used.

"Mills" fuel may be taken as an example of a commercial fuel, for it was first on the market, and will show the reader how he should mix his fuel with the correct proportion and type of ether. Other commercial fuel blends are sold with instructions on the container. These may slightly vary in the proportion of fuel to ether, but the principle is the same.

FUEL MIXTURES

(Shake the mixture well and keep corked.)

No. 1 *"Mills" Fuel (Blue Label)*:

Obtainable in small containers. Mix 1 part ether,

2 parts " Mills " special fuel. Shake well and keep the bottle corked. Ether is normal 0.720 " Ether Meth," or " Anaesthetic Ether," *obtainable from a chemist.* (Not commercial ether.)

Remember that " Mills " fuels must have ether added to them, and will not operate properly by themselves. They have the correct lubricating oil already mixed in the fuel.

The resulting mixture of fuel and ether should be perfectly clear. If it is cloudy, the wrong ether is being used.

Very minute diesels often run best on 1 measure " Mills " fuel, 1 measure ether.

No. 2 *The Simple Fuel* (" *My Mixture* ") : *If commercial fuels not obtainable.*

5 measures " Ether Meth," or " Anaesthetic Ether," *obtainable from a chemist* (see end of chapter *re* ether).

4 measures diesel oil, obtainable from a garage.

1 measure lubricating oil. This must be of a good motor-cycle air-cooled grade such as Castrol XXL.

Add 2 per cent. amyl-nitrate if obtainable.

Never use any thin machine oil, old sump oil, bicycle oil, or other substitutes as a lubricant.

(N.B. Assuming the measure is the same size.)

No. 3 *E.D. Fuel* uses a castor based lubricant, and is bottled with ether incorporated. It therefore merely requires the bottle being shaken. The container top must be kept on when not in use to preserve the highly volatile ether content.

No. 4 *Mercury Diesel Fuels*

Mercury No. 6, " All-in-One," requires no ether to be added, and does not have ether incorporated. It is very easy starting with great power output, but runs rather hotter than etherised fuels. It is therefore best to " run in " a new engine with etherised fuels, and No. 6 can be used when nicely run in. I seldom use any other fuel for all but the baby motors, because it saves so much trouble, and there is no ether to evaporate. It is rather too thick for my very small motors. *Mercury No.* 3 is blended with Essolube Racer lubricating oil and is a cool running fuel, which *must* have ether added. This may be from 1 part ether to 1 part fuel, or 2 parts fuel to 1 part ether according to requirements and maker's instructions. *Mercury No.* 8 is a ready-mixed castor lubricant based fuel for diesels.

No. 5 *Frog " Power Mix " fuel* is blended with " Aeroshell " oil, and must have ether added, in the proportion of 1 part ether to 2 parts fuel. (See page 101.)

No. 6 *A " Fixed Head " fuel as recommended by the " Micron " French manufacturers, (with extracts of their remarks in italics). Other " Fixed Head " manufacturers specify different fuel mixtures to suit the compression they use.*

" The motor has been designed to function with the following mixture :—

Paraffin (*medicinal*)	15%

N.B. Do not use ordinary lamp paraffin.

Lubricating oil	10%
Ether	75%

" This is the mixture which gives the best results from the point of view of both maximum power and ease of starting.

" If medicinal paraffin is not available you can use vaseline in the same proportions.

" If it is not possible for you to obtain either medicinal paraffin or vaseline, you can make use of the following mixture :—

Motor lubricating oil	20%
Ether	80%

" Easy starting is mainly dependent on the nature of the oil used. Results can vary *considerably* with different oils. For your first trials use more oil, 30 per cent. to 35 per cent. The adjustment of the fuel valve is less sensitive."

N.B.—A good alternative home-made mixture to No. 2 is that used by the "E.D." diesel. 1 measure each castor oil, ether, paraffin (commercial), or diesel oil. Mix castor oil with ether, than add paraffin or diesel oil. Shake well when mixing.

POINTS OF INTEREST

I may as well explain here that diesels are dirty to operate. They fling out a large proportion of the oily nature of the fuel from their exhaust ports. At the end of the day's flying it is desirable to clean up the engine and the plane or boat to prevent dust and grit collecting on the model. Diesel oil should be washed carefully from the hands after use, as it affects some skins adversely.

A diesel model often leaves an interesting trail of smoke behind it, which in the case of the model aeroplane sometimes gives a picture of the airflow behind the wing. I remember on one winter's day, one of those very calm frosty days, a high-wing model of mine was flying with the sun behind it, and I obtained the most perfect silhouette view of the model and the diesel smoke trail against the sun. I noticed that the down wash from the wing carried the smoke in a long sweeping curve down from the wing so that it missed the tailplane "by miles." This free flight test was far superior to any wind tunnel tests, which are often subject to tunnel wall interference, etc. The practical view I had, proved to me the fallacy of the so-called interference with the tailplane on a *slow flying* model of the high-wing type, provided the tail is mounted in line with the wing or only slightly below.

REMINDERS

1. Always keep your fuel bottle well corked to prevent the ether from evaporating and leaving the other constituents in a greater proportion. This does not apply to Mercury No. 6.
2. Keep the bottle away from fires and cigarettes. Also remember ether is not only dangerous because of its explosive and "flash-back" qualities, but also its anaes-

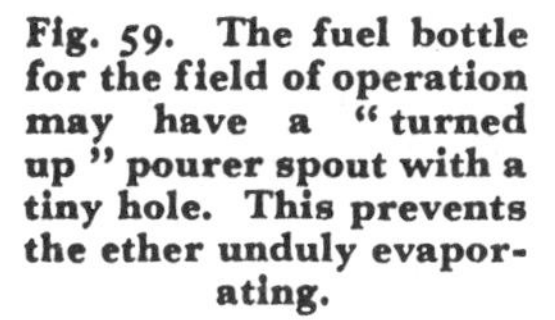

Fig. 59. The fuel bottle for the field of operation may have a "turned up" pourer spout with a tiny hole. This prevents the ether unduly evaporating.

thetic propensities. If you run the engine indoors, see that the room is adequately ventilated.

3. Carefully strain fuel when mixing to extract fluff and impurities.

A FUEL BOTTLE FOR OPERATION ON THE FIELD, AT THE POND-SIDE, OR RACE TRACK

An old "Dettol", graduated disinfectant bottle, or a graduated medicine bottle will make an excellent container, because it is simple to measure the proportions to be used, as one can see the fuel and the graduated markings.

A little filling funnel can be used and the fuel poured from the bottle into the tank *via* the funnel. *In this case the bottle must have its cork replaced every time to conserve the ether content.*

An even better plan dispenses with the cork, the filler spout being so very small that the ether does not appreciably evaporate, consists of a simple pouring spout of thin model aeroplane brass tube soldered into a metal bottle cap, as shown in Fig. 59. A tiny spout must be turned up from brass, as shown, with a very small hole. This brass spout is soldered to the brass pouring tube. A metal screw-on top for the bottle is obviously necessary to permit the brass tubes to be soldered into it.

The "Alton Valvespout" oil-can is obtainable at most ironmongers and has become very popular with modellers, because the spout has a quick screw-threaded stopper that seals the open end instantaneously without any danger of losing the top. The top remains *in situ* even when the spout is open to pour from its convenient-sized opening to suit the average model tank. When closed, the ether content of the fuel cannot evaporate.

FUEL TECHNICALITIES

Do not be frightened or put off the model diesel by long-winded discussions over fuel. Leave this to the genuine fuel technicians to sort out, and also to the armchair experts—the latter derive a lot of fun from their discourses but very little of practical value.

The fact that stands out is that the fuels that I have mentioned will run your engines perfectly satisfactorily, and that is what matters to the keen modeller who requires running results. In any case, one short chapter on the intricate matter of fuels could never cover what is a life's study of the fuel expert.

PRE-IGNITION

The term "pre-ignition" that is sometimes used in connection with a *model* diesel is rather inaccurate, because it refers to ignition in the petrol engine *before* the ignition spark takes place, due to varying reasons. In the full-sized diesel the ignition is definitely timed to take place when the fuel is injected.

In the model there is no absolute timed ignition point. The gas "explodes" when it is ready, due to the heat of compression.

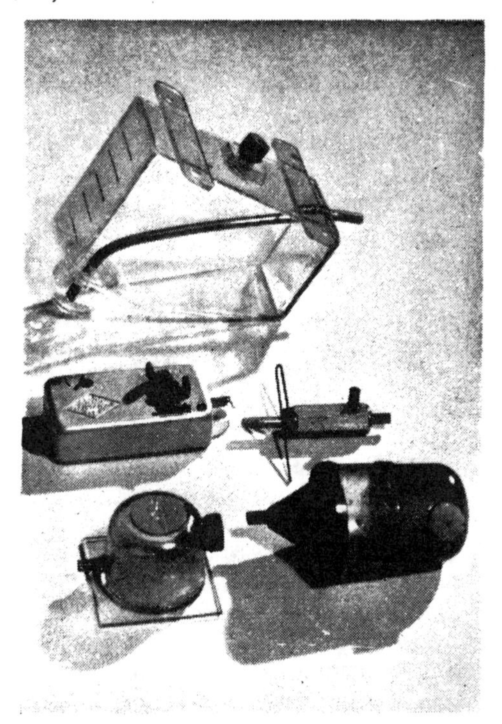

Fig. 59A. Useful Fuel Accessories. At top the large E.D. tank with graduated fuel level marks for radio flight of long duration. Note the well at lower corner to retain even flow when stunting. In middle, the E.D. clockwork timer to operate a switch on the F.G. fuel cut off seen beside it. Below two very useful plastic tanks for free flight and control line flying

There cannot, therefore, be pre-ignition in this sense. There can, however, be early or late ignition. There can also be detonation or combustion lag.

Detonation is produced by too rapid burning of the mixture and is caused by too weak a mixture or too high compression or both. Combustion lag is caused by too rich a mixture and its attendant slow combustion, together with a low compression, or an unsuitable fuel.

I have a sting in the tail to hand out lest I shall be accused by the technically-minded of making fuel matters sound too simple for Mr. Everymodeller, and therefore taking all the kick out of the problem.

There is no denying the fact that we may not have arrived at the ultimate fuel for model diesels. There are other accelerators besides ether which are suitable and which can be ready mixed.

As an instance of a non-etherised fuel, the new " Mercury No. 6 " diesel fuel has been blended by the Anglo-American Oil Company and starts very well. This fuel is composed of six components and literally does not function correctly if any one of the ingredients is not included. It has a lower fuel consumption than the normal etherised fuel, and I have found it start a diesel engine surprisingly easily, producing outstanding power. I use it a great deal because of the trouble saved in obtaining ether.

To prove how near we are to being able to run on normal diesel fuel oil by itself, I have tried a simple little experiment. I have filled my 3.5 c.c. B.M.P. engine tank with pure unadulterated diesel fuel oil. The engine will not start on this, but if one squirts a little of a normal etherised fuel into the induction pipe, the engine starts up, *and it then runs out the tank load of pure diesel fuel oil.* That makes one think! It means that once the engine is started it keeps warm enough to run on the pure diesel oil. The compression has to be raised a little and the fuel needle-valve opened a good deal more. The exhaust is rather smoky and the power is less, nevertheless the engine runs and swings its normal propeller at quite a good turn of speed.

I have also seen a " hot " square 4 c.c. model diesel engine fitted in a model car do the same but at a higher r.p.m. and giving greater power, probably due to the extra heat generated at the higher r.p.m.

For slow speed running on large diesels, a mixture of only 5 per cent. ether has been found to give starting and good running. A mixture of only 2 per cent. of ether will give good running if a little extra ether is dropped into the induction pipe to give a start.

Let us, therefore, for a few brief moments delve into the simple facts of fuels, without being too technical over names that may fog the issue.

ETHER

I have seen it stated that one should not use " methylated ether " in model diesels, or dire results due to the presence of the corrosive influence of sulphuric acid may occur, and that the modellist must therefore use only " anaesthetic ether." In actual fact, the doctor sometimes uses methylated ether and sometimes anaesthetic ether. *Both* " methylated ether " and " anaesthetic ether " are pure, and there should not be acids in them to burn out the vitals of your diesel.

A chemist and I carried out a litmus paper test of both ether meth and anaesthetic ether. Neither showed the slightest trace of acid, whereas a drop of sulphuric acid from another jar turned the blue litmus paper a lovely shade of pink. I think the misconception about ether meth has grown up because ether meth is a distillation of sulphuric acid. As a result, some people have jumped to the conclusion that there is acid remaining in ether meth. This is not so, provided it is the pure type.

There is, however, a third ether called " commercial ether," which may severely damage your engine by corrosion. That is why I recommended earlier in this chapter that you should go to a chemist to obtain your ether. There is practically no difference in price between methylated ether, or anaesthetic ether. The only difference is that methylated ether has a small tolerance allowed in its specific gravity from 0.720 to 0.750, whereas anaesthetic ether has to be 0.720. The diesel does not care which you use.

HIGH AND LOW OCTANES

There are many fearsome names in the fuel groups which we need not worry about unless we are becoming a director of a petroleum company, or its chief chemist, but we might perhaps mention that low octane fuels will cause detonation—which is the cause of engine knocking. Ethers and paraffins are usually low octane fuels. High octane fuels burn evenly under compression and do not readily detonate. They are therefore not suitable for model diesels as an only fuel.

The detonation of low octane fuels must, however, be controlled, because fuels for model diesels must not have too short or too long a combustion lag. In the latter case it may be that the fuel detonates, drives the piston down, and when it is coming up again, the piston receives the full force of the delayed explosion that has by then arrived at its maximum effort. No engine will stand much of that sort of thing. Too short a combustion lag on the other hand will obviously give violent detonation and knock the engine to pieces.

It is generally accepted that the ether in the fuel ignites and fires the diesel fuel oil, which with the lubricating oil tends to decrease the combustion speed of the ether. This is another way of saying they quieten down the violence of the detonation. The lubricating oil at the same time attends to its work of lubrication, and also acts as a further damper on the combustion speed.

According to the fuel experts, the addition of 1-2 per cent. of amyl-*nitrate* to diesel oil, noticeably increases the ease with which such a fuel will self-ignite in a diesel engine, but once combustion has started it has no significant effect upon its speed. In practice, I find the engine runs more smoothly. Do not use amyl-*nitrite*.

The low flashpoint of ether is incidental to its effect of increasing the ignitability of a diesel fuel in an engine. Petrol would also lower the flash point but would decrease the ignitability.

Ether combines high volatility with ignitability to a remarkable degree.

I have found in practice that paraffin and petrol are not so effective as diesel fuel oil. That is why I use diesel fuel oil in my simple mixture rather than the other two fuels that are sometimes recommended. Paraffin or petrol may cause misfiring, knocking, excessively smoky exhaust, and generally rough running.

A LOWER ETHER CONTENT

I have recently started up with complete regularity, and run with the same power, a certain 2-c.c. British diesel, using a 12 per cent. ether content and Mills diesel model fuel. It confirms my belief that the model diesel should be developed for power lawn mowers, baby outboard boat motors, and similar full-sized " utility machinery."

As this book goes to press, Frog " Power-Mix " has been placed on the market as an ether-less ready mixed fuel in a special metal container with screw-spout top. A test sample appeared to start easily and run well.

CHAPTER IV

HOW TO START DIESELS, INCLUDING A STARTING DRILL

GOOD STARTING MAKES POWER FLYING, BOATING AND CAR WORK A PLEASURE

BEFORE reading this chapter it is highly desirable to read the previous chapter on the matter of fuel.

Model flying is made or marred by easy or bad starting. This also applies to the good name of an engine. The general principles for starting also apply to boats and cars. I am including a few specialised notes on these at the end of this chapter.

The diesel is a very simple engine to start provided it is of good design and construction, and provided the individual attempting to start it understands certain principles and then applies these with patience in an orderly sequence.

Some time ago, I visited two model engine manufacturing firms, whose engines I had owned and flown, and which I knew to be good starters. I was shown their method of testing every engine that left the factory for good starting and power output. The test was rigorous and engines were sent back for attention by the tester if they did not start up in the specified time by hand-swinging a normal propeller as advised for flying.

Both these firms had a very excellent reputation, and yet there were certain individuals who could not start their products, although these were sent out with " controls set." In a few cases, people had written rude letters and said hard things about the engines. I was perfectly satisfied that these were quite unjustified and that the fault lay at the door of the complainers. Even the best manufacturers suffer these unjust complaints through lack of elementary knowledge on the part of certain of their customers.

One of the most difficult things that a maker has to face is to get all new owners of his engines to read the simple starting and operating instructions sent out with engines and then to abide by these instructions.

Some new purchasers are too eager to get at the engine, whilst others think they know better. There are people who think they know all there is to know about model engines because they own a car or a motor-cycle. There are also some who are just stupid, and fiddle until they lose the original settings. They cannot then find these settings because they have not marked them or made a note of them.

There are others who have had some engineering training, but fail to appreciate the fact that model engines, although fundamentally the same as full-sized engines, require a different technique of handling. Happily there are also many sensible people who read their instructions and use their common sense. It is my hope that this book will appeal to this type of person, and perhaps win over a few of the others.

Many people have come to me and said in various ways, " This engine is no good, I can't get it to start, and so I took it to Mr. So-and-So, who is a ' full-size ' engine expert, and diesels are his daily fare. He can't get it to go. I cannot think how a huge firm like Such-and-Such & Company can put a thing on the market that will not start. Robbing the poor, I call it." Or words to that effect. A few slight adjustments to undo the heavy hand of the " full-sized " engineering expert, who one sometimes suspects being an apprentice in a country garage, chiefly concerned in sweeping the shop up, followed by a few flicks of the propeller in the right way, and the poor manufacturer's reputation is vindicated through a burst of diesel power that surprises the owner, and sometimes almost annoys him that it should be so simple. The fact is that the modern manufacturers do not put dud engines on the market, and a new owner would be wise to suspect flaws in his own efforts rather than in the engine. This happy state of affairs was not so a few years ago at the beginning of model diesel popularity in Britain. Most of the duds have died a natural death.

One of the most common faults is to screw down the contra-piston on a diesel hard should it fail to start quickly. This faulty reasoning being "because it is a compression ignition engine, and therefore if one raises the compression really high the thing is sure to start." I can assure this type of individual that he is most unlikely to obtain a start because the compression will be too high, and he is most likely to damage his engine!

If we are novices, or even fairly experienced petrol modellers, let us READ THE INSTRUCTIONS AND ABIDE BY THEM. Let us mark or otherwise keep a record of the settings sent out by the makers, of needle-valve and contra-piston—we shall then have a basis on which to start off, and to refer to if we make a mess of things! It must, of course, be realised that the settings as sent out may vary a little due to a slightly different fuel and other reasons, but they do form a basis from which to work.

NECESSARY PRELIMINARIES

Make sure you have a correct propeller as recommended by the makers. Diesels and glow-plug engines are very much affected by the propeller size, balance and weight. The diesel, generally speaking, is a slower revving engine than the glow-plug engine, which is not of much use unless it can "rev." The propeller controls engine speed by diameter and pitch. Its design is therefore vitally important to get the best performance out of any particular engine type. For each type has its own peculiar characteristics. Maximum horse-power is produced at different revolutions.

Also make quite sure that the engine is firmly mounted and that mounting bolts are tight after each flight or run. This is very important, due to the high compression which causes rougher running than with petrol engines; and also the swinging of the propeller against high compression. Vibration-free running in the model is necessary; vibration also upsets fuel mixture strength.

It is most astonishing how many people fit incorrect propellers or mount their engines on some flimsy mounting, very often with the unfortunate engine lugs unevenly supported.

Make absolutely certain that you have the correct fuel mixture with the correct proportion of constituents as recommended by the makers. So many people try weird mixtures and curious lubricating oil substitutes, such as old car sump oil, machine oil, bicycle oil, and so on, often because " an engineering friend " has told them that, being a tiny engine, it requires thin oil. Do not listen to such false prophets and advisers. Unfortunately, I find that they are legion.

Be sure that your fuel is filtered, kept in a clean bottle and well corked so that the ether content does not evaporate, also see that you have a suitable pourer.

Check that the propeller is screwed well home and tight and if possible that a lock-nut is fitted. We should check this tightening of the propeller nuts after each run or flight, as some diesels have a habit of shedding propellers. Do not over-tighten with a vast spanner, as threads can be stripped.

The propeller should be bolted up so that it is coming on to compression just before the top of the starting swing.

THE NEEDLE-VALVE

On a model petrol engine the needle-valve setting is critical.

This is not so critical in the case of most diesels, but there is a definite relation between the needle-valve of the carburettor and the contra-piston—once the needle-valve setting has been found in the case of the diesel, very little alteration need be made, provided the constituents of the fuel are carefully adhered to each time a new supply is mixed up, and provided the fuel bottle is kept corked to prevent evaporation of ether.

We should therefore remember, " DO NOT FIDDLE WITH THE NEEDLE-VALVE."

Get the setting correct and leave it. Make a mark where the needle-valve should be set. You can then always refer to this mark. Should there be no setting given with the engine when it is purchased, or should the operator lose the setting, the drill given in " Procedure B " overleaf should be carried out.

When I used to go in for competition flying and hydroplane racing, I made a rigid practice of finding the best needle-valve setting, or the correct jet in a fixed jet carburettor before the event, and then keeping to this. *I never fiddled with this setting ;* as a result I was able to obtain quick starting and reliable performance, when many of the competitors were juggling with adjustments in a wild desire to get that little extra bit out of the engine. This so often results in confusion. Mr. Rankine, the famous hydroplane exponent, adopts a similar attitude, whilst Mr. Curwen, one of the best known car racers in this country, is also a believer in the principle. I will quote the words of a letter he recently wrote to me referring to his petrol race car. " Two or three pulls to choke, and the engine starts every time with the choke half closed. *I never alter the jet setting*—at least, I have not done so for the past four years or so on this engine. Speed has been very consistent for some time now."

On the other side of the picture, at the 1946 " Bowden International Power Trophy " for model aircraft, I have seldom seen so much trouble over the starting of engines, petrol and diesel. From careful observation, I should say that 80 per cent. of this was due to the competitors wildly playing with their needle-valve adjustments, as soon as their engine did not respond at once to the starting swing. The results were chaotic, for " the known best run settings " were lost, and engines became choked or starved, and were often sent off weak or rich in the heat and excitement of the competition fever, when and if they did get started. Almost everyone used their full time limit of three minutes and three false starts which delayed the whole competition. Most of this would not have occurred if the competitors had left their needle-valve settings at the " known best run position," and then choked or doped to suit their particular engine to obtain the start from cold, always provided, of course, that the needle-valve orifice is kept clean.

HOW TO SWING THE PROPELLER

I have watched a considerable number of people who say that they cannot start their diesels, and petrol engines too, and I have noticed that one of the real stumbling blocks is the swinging of the propeller. Some get quite testy if they are told they are swinging improperly. They are surprised when one gives two or three swings to their "difficult" engine, and off it goes.

There is a right and a wrong way of swinging a propeller. Let us first see how not to do it, the way many newcomers

Fig. 60. The starter cramps his style by not facing the propeller properly.

try to do the job. Then let us see how to do it. I have taken a few photographs which I think may help. The methods I advocate are my own and not necessarily the only way, but I can at least recommend them as a way of getting a start. Goodness only knows how many propellers and flywheels I have swung in my life-time!

(1) *The Wrong Way*

Look at Fig. 60 and you will notice that the starter is bending over his engine from the half rear, or side, where he cannot really get at it to swing lustily.

Notice how he pulls up on the propeller and therefore

cannot flick it over the top and downwards with a good three-quarter follow through at speed.

He has his swinging finger too far out on the propeller blade. This makes a long arc for him to swing and so reduces the speed of the "flick" over compression, and the speed with which the propeller rotates and the engine sucks in and compresses the mixture. See Fig. 61. It should be realised that *in order to obtain the necessary heat from quick compression to fire the gases the speed of the swing or flick must be great. A weak, half-hearted swing is useless!* Fear of a kick-back sometimes causes this. A weak swing actually encourages a kick-back!

Fig. 61. The starting fingers are too low down the propeller.

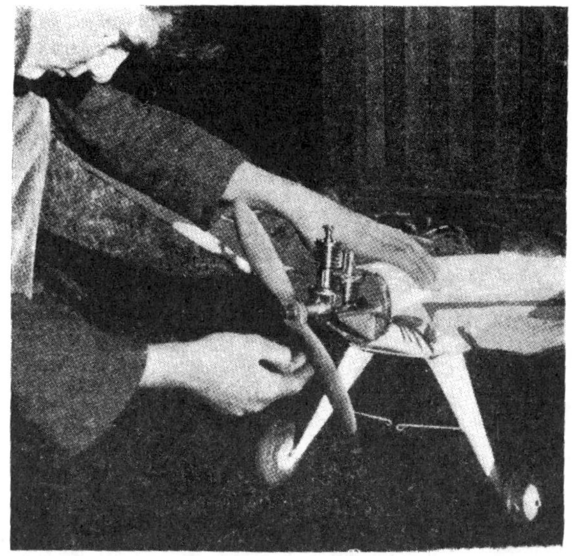

Fig. 62. It is better to place the fingers near the centre of the propeller. This allows a quick flick.

(2) *The Correct Way*

Now look at Fig. 62.

It will be observed that the starter is now *in front of* the engine where he can swing *with a follow through.*

This follow through is most important and the lack of it is the reason for a great many failures to obtain a start. As soon as people are shown how to obtain it, they start their engines. It is like the follow through wrist flick of the golf swing. It adds power and speed to the stroke.

The starting finger of many aeromodellers leaves the propeller when it has swung the propeller for only half a revolution, and in some cases less. This is sometimes due to being afraid of the engine and sometimes due to lack of practice. The follow through flick is a technique that has to be acquired, and the propeller must travel through at least three-quarters of the circle.

The swinger must not be afraid of the engine! The starting finger should be placed about halfway down the propeller. This permits the operator to flick the propeller over more quickly than when the finger is placed at the propeller tip.

Position the propeller correctly. In order to be able to swing the engine properly over compression to ensure easy starting, it is essential for the propeller to be attached to the crankshaft in the correct position relative to the compression point at the top of the piston stroke. Therefore screw up your propeller on the crankshaft so that it is coming on to compression just before the top of the starting swing.

A STARTING DRILL

If we have our preliminaries as mentioned correct, and we have mastered the starting swing as detailed above, we can best obtain success by practising a simple drill. This gives us an orderly sequence. Below I give a simple starting drill (a) *when the makers' settings are known.* This is followed by starting procedure (b), assuming both needle-valve and contra-piston settings have been lost, and (c) for fixed-head engines.

(a) *Starting drill when the settings of needle-valve and contra-piston are known*:—

When the engine is received from the makers, the settings should be noted and marked, so that they are not lost. They then form a basis from which to work.

(1) Fill the tank. (Never leave fuel for long periods in the tank, as the ether evaporates and leaves an incorrect fuel mixture.)

(2) Check that known settings are set as indicated, and make sure that the flight timing device is open.

(3) Increase compression by screwing down compression lever a quarter turn from "running setting," i.e. raise compression slightly. For exceptions, see remarks at the end of this drill.

Fig. 63. Here we see a well-known competitor facing his 1 c.c. "Frog" diesel and starting easily in "The Bowden International Trophy," 1946.

(4) Suck in fuel by swinging propeller smartly two or three times with finger over intake (refer back to Fig. 24, Chapter I, if in doubt of how to do this). Wet fuel on finger shows that fuel is being sucked up. Diesels take a lot of fuel to start. Do not overdo however—gravity-fed engines flood easily.

(5) Swing propeller smartly and when engine fires return contra-piston to original "running setting."

(6) If engine gets "over hard" and will not start as described overleaf in Procedure B, para. (7), unscrew contra-piston one turn—"obtain relief," then screw up contra-piston to "start" position, and swing to start.

(7) Finally, make slight adjustments to obtain maximum power. *Do not over fiddle with adjustments*, as diesels are not as critical as petrol engines—once a sound adjustment is obtained, they run steadily until the fuel is exhausted.

(8) If your model is one that climbs at a steep angle, test

the running fuel adjustments to suit by holding model with nose up. See Fig. 70.

A good diesel will be found very reliable and extremely simple to start. It is all a matter of the correct fuel, the correct needle-valve setting, and the correct compression. *There is no worry about the electrical spark*, for it does not exist. The fuel mixture is ignited by the heat generated by the high compression. As the engine warms up, we obtain an extra source of heat. The fuel will therefore be ready to ignite at a lower compression adjustment. We therefore reduce the compression slightly, as in para. (5) above.

If we do not reduce the compression slightly when the engine is warm, the fuel will ignite too early. The warning symptoms are—at first the engine note sounds less free, and later, the engine will gradually lose speed and sometimes even stop altogether. This puts extra stress upon the engine. *Keep the compression as low as possible for a long engine life.* It may sound unnecessary to say that there is no spark. I have however, had several letters from enthusiastic new modellers saying they cannot get a spark from their diesels! No names, no pack drill.

With regard to the exceptions mentioned in para. (3) above: there are a few diesels on the market which I have found to require no alteration of the contra-piston between starting position and running position. There are also a few engines which start very " wet " which actually like the C.-P. to be slacked off slightly to start, and then screwed up. The advantage of the contra-piston lever in these cases is that slight alteration can be made to compression ratio should one change the fuel in any way, or should one flood the engine by mistake.

DIESEL KNOCK

This phenomenon is caused by shock from the whole structure of the engine, and should not be confused with local pinking of a petrol engine.

Diesel knock is caused on models by too great compression, as explained above and also by a too weak mixture combined with over compression.

(b) *Procedure, assuming both needle-valve and contra-piston settings have been lost.*

(1) Unscrew contra-piston lever completely. Take right out if there is no stop fitted. If stop fitted, unscrew to stop.

(2) Shut needle-valve.

(3) Turn propeller over smartly, which should raise contra-piston to the top of the cylinder. This usually causes a smart crack to be heard, as though something has fractured. Actually, no damage has been done, but the contra-piston has shot up to the top of the cylinder through the force of compression.

(4) Replace C.-P. (contra-piston) lever, and screw down gently until it meets resistance of the contra-piston itself, or if stop is fitted keep at " unscrewed position."

(5) As no setting is known for needle-valve, it can be found only by experiment, and it is fairly safe to try one turn open. (Some engines require as many as four turns open. This depends largely whether the needle-valve is a fine or coarse taper.)

Try one turn—choke engine by placing finger over the intake and swing propeller smartly twice.

On removing finger, and again swinging, a wet sucking noise is heard if fuel is being sucked up, also fuel can generally be seen on the finger-tip which has been withdrawn. If no sign of fuel is present as above, repeat process, with needle-valve open another turn. If still no result, try another turn.

CHECK UP THAT THERE IS FUEL IN THE TANK !

(6) Fuel is now assumed to be present—swinging the propeller as smartly as possible with the right hand, very gradually screw down the contra-piston with the left hand—when the correct setting has been reached, the engine should begin to fire.

If the needle-valve setting is too weak, engine will run and then stop, and no further turning of the propeller will restart it—open needle-valve approximately a further half turn. On the other hand, *if the setting is too rich, the engine will run with regular bursts of speed up and down the scale.* This indicates that it needs either more compression (C.-P. lever screwed down) or weaker mixture (needle-valve screwed down).

It will be appreciated that it is not possible to give a newcomer to diesel work a completely foolproof explanation of starting when all settings have been lost. Some experiment and " starting sense " is often required, and this can be obtained only by practice *along the right lines*, as laid down above.

(7) *The " over-hard " engine due to excess of fuel.*

One snag must be clearly understood, and be again emphasised—*the compression space, i.e. the space between contra-piston and the piston at top dead centre in diesel motors is very small (often only 1/32 in.) in order to obtain the necessary high compression ratio to create auto-ignition.* It is therefore possible, particularly on the smaller motors, to suck in too much mixture in which the oil content remains, whilst the ether may evaporate—oil is practically *incompressible.* This will make the propeller either very difficult or impossible to get over top dead centre. If the operator attempts to force the engine over top dead centre, he must be prepared to bend or fracture something inside the motor. It is then useless to complain to the manufacturer about such a motor. He knows the reason for the damage—you !

Slight difficulty in turning the engine over compression is perfectly correct, and in fact necessary to start. But there should be no difficulty in recognising when the engine becomes really " over-hard " and the operator is forcing it to breaking point.

Should this " over-hard " condition occur, i.e. when too much fuel has been sucked in, the *contra-piston lever must be unscrewed half to one turn.* The engine is then swung over compression, which raises the contra-piston and creates relief. After relief the compression can be slowly raised again by screwing down the C.-P. lever whilst swinging the engine, until a start occurs. As one becomes experienced it is quite possible to " sense " an engine with too great a compression to start, although it has not reached the " over-hard " state, which is often known as " hydraulicing."

THE " HOT " HIGH SPEED MOTOR

During the past year or so a number of very high efficiency diesels have been produced for speed competition events. These are sometimes best started by dropping two or three drops of fuel into the open exhaust port from the fuel can's spout, and swinging with vigour to start. The normal running settings are left. A stuck-up contra-piston often worries new-comers to the diesel. This is explained at the end of this chapter.

(*e*) *Starting engines with " fixed heads," i.e. the engine with no adjustable contra-piston.*

Manufacturers who make engines which are not fitted with the adjustable piston, issue instructions for a fuel mixture. The proportions and ingredients of this mixture must obviously be adhered to because deviations in the fuel mixture cannot be compensated by compression adjustment through a contra-piston.

Should the " over-hard " engine be experienced, as has already been explained, due to too much fuel being sucked into the cylinder, the only rapid cure is to turn the engine backward until the exhaust port is seen to open. Now turn over on its side and drain out fuel from cylinder—then commence starting operations again.

I am giving below an abridged extract of the very excellent starting instructions drawn up by the French " Micron " diesel makers. The 5-c.c. motor, an example of which I have (see Fig. 6), has a fixed head and can be taken as a very sound fixed head design.

If the reader will refer to Chapter III, he will see the

recommended fuel by the " Micron " firm ; he can then read how to start the " Micron " fixed-head engine below. These instructions will suit many other fixed-head engines where the induction pipe points downward from a crankshaft rotary valve, as in the case of the French " Micron " and the British " Owat."

Extracts from " Micron " instructions

" Easy starting is mainly dependent on the nature of the oil used. Results can vary *considerably* with different oils. If you encounter difficult starting, change the oil content. For your first trials use more oil, 30 per cent. to 35 per cent. The adjustment of the fuel valve is then less sensitive.

Starting

(1) Close the fuel valve. (Needle screwed right home.)

(2) Fill the tank with fresh fuel.

(3) Set the flight timer control plunger by pulling it upwards until the catch engages, i.e. fuel is allowed to flow.

(4) Open the fuel valve three or four turns. The mixture will run of its own accord and will fall drop by drop from the air intake. If the mixture does not run in this manner (owing to an obstruction), close the air intake opening with the finger and give the propeller one turn.

(5) As soon as the mixture is dripping from the intake pipe, close the fuel valve completely. The mixture, having filled the intake pipe, produces an excessively rich mixture of gas to be drawn into the engine which prevents an explosion.

(6) Swing the propeller (anti-clockwise) until the motor fires.

(7) Now open the fuel valve approximately $\frac{1}{4}$ or $\frac{1}{2}$ a turn and continue swinging the propeller.

(8) Two conditions may present themselves :—

(*a*) *The fuel valve may be too far open.* The explosions cease and the compression becomes " stiff " to pass. The mixture flows drop by drop from the intake pipe. (It is gravity feed on these engines unless they are inverted.) Close the fuel valve and swing the propeller until the engine starts a series of explosions of greater and greater length until it runs without stopping.

(*b*) *The fuel valve is not sufficiently open.* The motor fires but knocks at compression—open *very slightly* the fuel valve and swing the propeller until the engine starts a series of explosions of greater and greater length until it runs without stopping.

(9) As soon as the motor runs, open up the fuel valve and search for the setting which gives optimum results where the motor does not knock.

After a few starts, the knack of cold starting will be acquired. When warm, after the motor has run a little, starting will be instantaneous and the regulation of the fuel valve less sensitive.

Too weak a mixture causes knocking and puffs of smoke from the exhaust.

A flimsy engine mount causes air bubbles in the fuel due to vibration. This creates uneven running."

Should you own a *fixed-head diesel* with the induction pipe upwards or horizontal and where the fuel is sucked up and not gravity fed, the starting instructions are very similar except that the owner cannot watch for the dripping of fuel. The fuel must be sucked in by choking by the finger, but you must be careful not to suck in too much, as there is no contra-piston to unscrew momentarily and give the engine relief for excess of fuel.

The reader has been given three different situations to choose from which have been explained in detail.

First: The drill for starting with a diesel fitted with a contra-piston and where he knows the maker's settings of contra-piston and needle-valve.

Second: The method of finding the settings if they have been lost for an engine fitted with a contra-piston.

Third: A method of starting a fixed-head (i.e. no contra-piston) diesel.

These three situations should cover any diesel starting problem likely to be met with. The main thing being to understand how the diesel works (described in Chapter I) and then to practise starting on sound lines until a "starting sense" is gained. The midget type is dealt with below.

Although it has taken a long time to write about the matter, in order to cover all possible details and troubles, in actual fact starting is quite simple and is soon mastered. If the reader has any form of trouble, I feel convinced he will find the answer somewhere in this chapter.

STARTING THE MIDGET TYPE

The midget type of diesel is not as easy to start as the larger type from 1 c.c. upwards, owing to the small clearances in the cylinder-head, i.e., between contra-piston and piston.

The engine often sucks in neat fuel when choked. The very small space in the cylinder-head becomes full of fuel, and it is therefore absolutely necessary when the engine becomes "hard" to slack back the contra-piston a half to one turn, according to the feel of hardness. When the engine becomes free, *slowly* return contra-piston to the running "setting" whilst swinging the engine.

During this process the engine will most probably begin firing. It is then that experience will show precisely the way to manipulate the contra-piston. It should never be necessary to touch the needle-valve setting, except for minute adjustment for maximum revolutions. Also remember that the contra-piston adjustment must be very delicate because of the very small compression space in the cylinder-head.

We thus have in effect only one adjustment to make for the midget engines, namely, the contra-piston. This simplifies the problem. It cannot be emphasised sufficiently that the least possible fiddling with adjustments, *combined with patience*, obtains results. It is most noticeable at competitions and elsewhere that starting difficulty is usually due to over-adjustment in an effort to get the motor to run before the mixture is properly vaporised and warmed up through the heat generated by turning the tiny motor over compression a sufficient number of times very rapidly.

Once a diesel, particularly a midget, has started on its first run of the day, subsequent starts are more easily made, and it is frequently the best plan to leave all settings at the best running position, suck in twice and swing until a start is made.

I find that "Mills" fuel is particularly efficacious for my "Mite" engines of 0.7 c.c. once I have obtained the *exact* setting of the compression. I use ½ Mills-½ ether for midget engines because this greater content of ether does not clog up the tiny fuel space around the needle-valve.

There are exceptions to every rule! When the "Frog" 1-c.c. engine is inverted it becomes gravity feed, due to the tank fitting arrangements. When the engine is run upright the feed is suction. In the gravity-fed inverted position, care must be taken that the engine does not become flooded at the start. It is good practice to suck in one or two swings with the needle-valve open to the normal run position. *Then close the needle-valve, start, and when the engine produces a burst of song, open up the needle-valve to the normal best run position*, approx. half one turn open. With the engine in the upright position, and its suction feed, there is naturally no need for these special precautions. A small sleeve is fitted into the induction pipe for upright running of the "Frog 100" This improves starting in the upright position, but should not be used when inverted.

STARTING BOATS AND CARS

In this chapter, we have discussed how to swing a propeller for aero-engines. Many people will also want to use diesels in boats and cars where a flywheel is fitted.

In Chapter VI, I describe a car that is not fitted with the normal flywheel and can be started by a flick of one wheel, the drive then being taken up by a centrifugal clutch. However, the bulk of modellers will doubtless use flywheels for cars, and will certainly do so in boats. In the case of cars, a centrifugal clutch will most probably be fitted to take up the drive evenly as the car is sent away.

I have done a great deal of petrol hydroplane work, and I have built a number of V-type planing speedboats during which I have found that the most simple and reliable method of starting engines is to turn a groove in the outside periphery of the flywheel, where the leverage is greatest (not at the centre, where starting pulleys are sometimes fitted), and to use a round leather sewing-machine belt. The belt is held in each hand, as shown in Fig. 64. The hull or the car is either held down by a helper, or temporarily held to its starting-stand-cum-tool-box by thick elastic bands if the operator is on his own. By pulling the belt up from one side to the other smartly, and retaining just sufficient pressure for the comparatively large frictional surface of the leather to grip the V-eed groove of the flywheel, the engine can be turned over quickly for a couple to three revolutions, and a start easily obtained.

This simple method is very effective and cuts out special

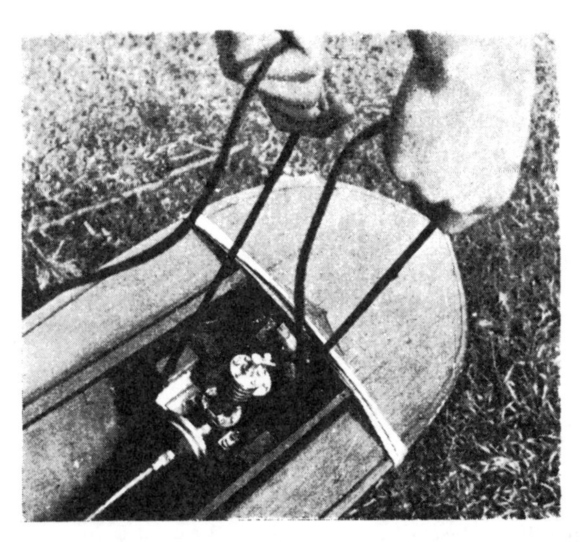

Fig. 64. Starting a boat engine is best done by the simple leather thong method.

Fig. 65. Mr. Curwen, the well-known car exponent starting up his car (in this instance a petrol engine) by a blind cord. He is noted for his quick starts, and has set up officially observed speeds of over 60 m.p.h. with his diesel car.

ratchet pulleys and other complicated mechanisms that are no more efficient for the main object of starting the engine, but do cause considerably more trouble to make and maintain. In a few cases car starter motors have been adopted with a friction drive to the model flywheel. This method, however, means the carrying of weighty starter motor and storage battery and is more useful for a club.

When starting by oneself, it is found impossible to pull up the thong using both hands, and at the same time to choke the carburettor by finger. I overcome this difficulty by doping the carburettor with two or three drops of fuel, or by fitting a little cork plug which I take out after two or three suck-ins. The cork is not lost, as it is on a string.

A description of diesel cars and boats is given at the end of Chapter VI.

Let us quote Mr. Curwen in his own words contained in a letter I have beside me, " The technique (or drill) is to hold the car down by kneeling on a board fitted with the hooks which register with a cross-member fitted across the bottom of the chassis below the starting pulley, the engine being started in the usual 'diabolo' fashion with a leather cord. Two or three pulls to choke and the engine starts every time with the choke half closed."

FINALLY—TOO HIGH COMPRESSION PREVENTS STARTING

The above is a fact so often overlooked by operators and even designers. If too high a compression is allied to an engine that is " sticky " to turn, then there will never be a start!

If the engine starts and then bounces back and forth, the compression is too high for the amount of fuel in the crankcase. Ease compression slightly and restart.

A STARTING TIP

If a model diesel's compression is not first-class due to wear or imperfect fit of the piston, provided this is not excessive, two or three drops of lubricating oil dropped into the exhaust port before starting will often help to obtain a start.

A STICKING CONTRA-PISTON

The contra-piston sometimes refuses to rise when the lever is unscrewed. This is due to the contra-piston being a very tight fit. A contra-piston has to be a close fit to be effective.

There are two methods of dealing with this trouble:—

(a) Clear out the excess fuel from the cylinder head by turning engine upside down and dribbling fuel out of *open* exhaust port. The engine may then start and the contra-piston go up as the metal heats up and expands.

(b) If the above does not prove effective, fill the cylinder head with oil through the exhaust port and by rotating the engine *gently* force the piston to compress the oil. *Be very careful because oil is virtually incompressible and the engine may be damaged if over forced.* If the contra-piston still remains stuck after reasonable pressure, it will have to be dismantled and the contra-piston eased in the bore. For the average person it is best to have this attended to by the makers.

CHAPTER V

DIESEL OPERATION—PROPELLERS—PROBLEMS—THE TIMING DEVICE—TROUBLE FINDING

THIS chapter contains a certain amount of repetition, as a very great deal of the subject has naturally been covered in the chapter on starting, for they are inseparable. I originally considered the advisability of combining the two subjects of starting and operating together in one chapter ; however, I finally decided to keep Chapter V as a separate entity because it would then perhaps form a summary and a last-minute method of impressing on the new modeller certain things that he should *not* do, in addition to those items that help towards sound management of the engine and the model to which the engine is fitted.

A certain amount of repetition is therefore inevitably combined with new information.

COMPRESSION RATIO PRECAUTIONS

It will be quite evident from the previous chapters that the compression ratio looms up very large in the model diesel's make-up and *life*. The word " life " is the one I wish to lay emphasis upon. If the owner wants his diesel to have a long and prosperous life, he must be careful *not to run the engine with too high a compression.* It is rather reminiscent of the man who cannot resist the extra few whiskies and sodas daily, with the result that his blood compression is raised and his useful life becomes shorter than it might have been !

Too high a compression knocks bearings to pieces, and it is also well to remember that it slows up the engine because it has to overcome excessive compression at the latter part of the compression stroke during the process of compressing the gases. It is like putting a brake upon the engine.

The important point always to remember is that once a start has been obtained (if a contra-piston is fitted), the contra-piston adjustment should be slacked back slightly until the engine runs best. There are a few engines that will start and run on the same compression ratio once the best setting has been found.

Also make sure that the mixture is not too weak. As stated before, too weak a mixture, and particularly too weak a mixture allied to too high compression, will cause what is known as diesel knock. The knock, in the case of the diesel engine, according to Mr. Ricardo, the expert on full-sized engines, is a shock wave coming from the whole structure of the engine, and not from a localised stress such as that which is known as " pinking " of a petrol engine.

I think it is worth remembering the reason for diesel knock, because the owner will then be *eager to stop this undue stress coming from the whole structure of his engine.* It is quite easy to stop the knock once the causes are recognised.

If a diesel eases up and gradually dies away in its power with a dead-sounding note, you may be sure that the probable cause is that you are running on too high a compression.

The gas is being expanded due to the heat generated by the running of the engine, more than when it was sucked in cold at the start. This is automatically raising the compression. *The answer is to let the engine run for a little while before releasing the model.* If it knocks, check up the compression when thoroughly warm, then reduce the compression slightly. The non-contra-piston engine of course cannot have this adjustment made, but a slightly richer mixture may alleviate matters by making the engine run cooler.

A number of purchasers of a certain diesel noted for its reliability and good starting have written to the manufacturer complaining that their engine will only run backwards. The fault in this case is that of the purchaser, because he is running his engine at too high a compression by screwing down the contra-piston too far. Another engine manufacturer I know has been cursed because his engine bearings wear out. This is due to the same cause, for I have found his engines wear very well !

Whilst on the subject of compression, I generally make a

practice of starting up any engine fitted with an air propeller, armed with a stout leather glove on my starting hand. It saves many an unpleasant rap over the knuckles!

If the compression is a little too high when starting, diesels have a habit of sometimes firing and then bouncing backwards on the next compression before one gets the starting finger out of the way. Hence the rap. Slightly slack back the contra-piston and have another try, fortified by the knowledge that you have a well-armoured and shock-absorbing finger due to the stout leather glove. *Bouncing back and forth means slightly too high compression.* Unscrew the compression lever slightly and restart.

I have seen unkind people fit mighty disc flywheels to run in their diesels. This is unfair on the diesel, because there is so much leverage and momentum that it can be easily forced over the " over-hard " compression that I have described, due to too much liquid fuel being compressed. Owing to the leverage and inertia of the large disc flywheel, the owner does not feel the very hard condition. Things then collapse inside the unhappy little diesel and the manufacturer gets the undeserved blame.

It is probable that you have read the chapter on starting and seen my advice in the case of a contra-piston engine, to " give the engine relief " by unscrewing the contra-piston lever one or two turns should you get too much fuel in the cylinder-head and obtain an " over-hard " engine. Maybe you own an engine with two stops fitted to prevent the lever being turned more than three-quarters of a turn. There are several engines of this type on the market, in order to prevent novices losing the approximately correct setting. In these cases all that can be done is to unscrew as far as the stop, and if the engine will not then turn over easily, resort must be made to the old dodge I mentioned in Chapter III, of turning the piston back, so that the excess fuel will dribble out of the now open exhaust port when the engine is placed on its side. Then turn, and if free, put the compression lever back to its normal start position. See Fig. 66. In the case of inverted engines, it will, of course, be necessary to up-end them to allow the fuel to drain from the exhaust ports.

Although a completely free contra-piston adjustment is

Fig. 66. The compression adjusting levers can be seen with special stops to limit movement on these two diesels.

the ideal for the knowledgeable man, we are seeing more and more engines fitted with contra-piston limiting stops in order to prevent the complete novice from losing all adjustment, or from grossly over-screwing down the contra-piston lever, and wrecking the engine by over-compression on a grand scale! The " Frog " diesels have a simple spring washer between head and adjusting lever which is most effective, for it prevents damage and yet permits of a considerable range of adjustment.

THE DANGERS OF ETHER

Don't smoke near an open bottle of fuel is another point which cannot be over stressed. Ether is very volatile and will flash back easily, and a nasty explosion may take place. If you are one of those uncontrolled chain-smokers, you will just have to stiffen up your controlling mechanism and take this warning seriously. Just because you have got away with it on a number of occasions does not mean that you have some special dispensation.

I need scarcely remark upon the fact that ether is used as an anaesthetic. So if I feel more dopey than usual I suspect the ether and do not think my normal powers are failing early in life. Keep the room well ventilated if running-in an engine on the bench.

THE MOUNTING, A REMINDER

I feel I must remind readers of the very great operational importance of fitting the diesel engine to a really sound and rigid type of mounting if reliable running is to be obtained. This can be rigid and yet of the detachable knock-off type, as already described, if desired.

In spite of many words on the subject, people still fit model engines to the most ridiculously flimsy pieces of wood, thin bits of plywood that wave about in the breeze, and other curious thin metal contraptions that cause a peck of trouble. I have described what happens earlier in this book, and I now end with a reminder to get down to tackling the really vital business of mounting the engine properly, both on the running test bed and in the model. *You will have no lasting success if you do not*!

THE PROPELLER AND THE LUBRICANT

When summing up operating difficulties which I have seen new modellers troubled by, I think that the fitting of incorrect propellers and the use of incorrect lubricating oil account for the two main operational troubles experienced, after the poor engine mounting.

As in the case of the mounting, I have given full reasons earlier in this book and need not recapitulate all these, other than to say that if the aeromodeller does not fit a correct and properly-balanced propeller and use the correct oil as recommended, then the ensuing trouble will be on his own head, and trouble will be sure to overtake him!

The "unbreakable" plastic propeller with flexible blades has now become available in Britain. There is a full range available from Keilkraft, Frog and E.D.

They are just the right weight for diesel engines and practically crashproof. The aerofoil section is kept correct by the flexible blades being flung outward through centrifugal force. I seldom use any other type for general purpose flying and radio flying.

If a propeller from the makers is unobtainable, the following approximate table will help to prevent a newcomer being given a totally unsuitable propeller. It can only be approximate because engines vary in power and r.p.m., and different models fly at varying speeds.

Remember that the propeller must not only be of the right diameter. It must be a heavyish one, and it must be balanced. It is quite easy to see if a propeller is balanced—put it on a nail or a piece of wire and spin gently by hand. If one blade always stops in one position at the bottom, then that blade is too heavy and the propeller is out of balance. If this is slight, a little sand-papering of the heavy blade will correct. The blades should stop at different positions each time the propeller is spun. There is nothing peculiar in a diesel propeller from that for a petrol engine, except that the diameter, weight and pitch must suit the diesel. A good diesel swings a larger diameter propeller than a similar capacity petrol engine, or glow plug engine.

A low-pitched propeller is the most suitable for a general purpose slow-flying aeroplane, and a higher pitch for a fast control-line model. The latter is suited by an 8-in. pitch for average requirements, in which case a slightly smaller diameter is required. A speed model requires an even higher pitch of approx. 10 in. to 14 in.

DIESEL ENGINES
(Approx.) Guide to Select Propellers

C.C. of Diesel	Diam. of Prop.	Pitch		
		F/Flight	Control Line	Speed
0.7	7 to 8 in.	4 in.	6 in.	Special to suit speed 9 to 14 in. approx.
1.0	7¾ to 9 in.	4¾ to 5 in.	6 in.	
2.0	9½ in.	5 to 5½ in.	6 to 8 in.	
2.5	9½ to 10 in.	5 to 6 in.	6 to 8 in.	
3.5	9½ to 11 in.	5 to 6 in.	6 to 8 in.	
5.0	12 to 14 in.	5 to 7 in.	8 in.	

PROPELLERS MUST BE BALANCED ! !
The higher revving engines should have the smaller diameters mentioned above.

Fig. 67. Flexible plastic propellers which are almost unbreakable were a valuable development of 1949. Here we note a plastic prop on a "Bee" 1 c.c. diesel fitted to the author's small 45 in. low wing model.

I have used slightly larger and slightly smaller propellers on all the above sized engines, but have found from experience that the sizes shown are the most satisfactory for general purpose slow-flying models.

SQUARE-TIPPED PROPELLERS AND THREE-BLADERS

For some little time now I have been using square tips to the blades of my petrol and diesel engine propellers, like the "paddle props." of many full-sized modern machines. I first tried this idea when I wanted to reduce the height at which I positioned my engine above the hull of a flying-boat. I made certain static thrust tests with an "Ohlsson 23" engine, upon which I was running a $10\frac{3}{4}$-in. propeller of fine pitch but normal blade area. I made a propeller of only $8\frac{1}{2}$ in. diameter for the engine and gave it square tips, but with considerably wider blades, so that it looked rather like a fan. I was surprised to find that, in spite of the very large reduction in diameter, the loss in thrust was only about 2 oz.—I had well gained this by reduction of frontal area for my flying-boat.

The same would apply to shorter undercarriages on land

Fig. 68. The stages of propeller carving.

machines. In the case of the flying-boat, I had been able to locate my thrust line in a more efficient place lower down and nearer to the centre of drag, which was a further advantage from the practical flying angle.

I then tried a less drastic cut of an 11-in. propeller as fitted to a 2.2. c.c. diesel. In this case, I squared off the end, which brought the diameter down to 10 in. and a little wider blade. The climb on the model was improved. It appears that the normal elongated tips are of little thrust value and they only cause unwanted drag.

I think the important point to remember is not to overdo matters so that the engine revolutions are unduly raised. Dr. Thomas has some interesting remarks to make on three-bladed propellers for scale models. See Chapter VI where a model of his is described.

AEROPLANE PROPELLER CARVING

The photograph of the different stages of propeller carving (see Fig. 68) should put readers on the right track if they wish to produce for themselves replicas of the original propeller they buy as recommended by the engine manufacturer.

Fig. 68. Stages in propeller carving :—

1. Mark out the top of the blank which must be of the correct depth to obtain the right pitch. Reduce ends to reduce tip angles.
2. Cut around the outline.
3. Cut away the inside faces.
4. Cut away the curved top surfaces of the "aerofoils."
5. Sandpaper smooth, balance, and varnish if desired.

MODEL BOAT PROPELLERS

I found that my speedboat "Flying Fish," described in Chapter VI, would run a 2 in. diameter propeller driven by a 4.5-c.c. petrol engine, whilst a 2-c.c. diesel liked a 1½ in. diameter three-bladed propeller. The blades being in each case ¾ in. maximum width and set at an angle of about 45 deg. with the tips given a little washout. A two-bladed propeller is just about as effective for normal speedy running. The three-blader had approximately the same area as the larger diameter two-blader.

These details form a useful guide for speedboat men. My tiny little speedboat "Sea Swallow," also shown in Chapter VI, and fitted with one of the 0.7 c.c. Mills diesels or a 0.8 c.c. "Amco" diesel, runs a two-bladed propeller of ⅞ in. diameter with blades ⅜ in. maximum width, a really midget affair which buzzes around at high r.p.m., approx. 7,000. This same hull when fitted with a 1 c.c. "Frog" diesel takes a propeller of 1 in. diameter, and 5/16 in. blade width. The E.D. "Bee" 1 c.c. diesel also uses a 1 in. diameter propeller.

A BOY'S TROUBLES

I gave my 16-year old boy a diesel, and he built a model aeroplane which I designed for him. It was his first power-driven model and I was interested to see how he would handle the situation.

Like many other boys whom I have watched, including those of between 40 and 50, he tried to swing the propeller incorrectly, as I have already described. Having shown him the

Fig. 69. A simple model designed by the author and built by his son. It is powered by a 2 c.c. diesel engine.

correct way, he at once got over the starting difficulty in exactly the same way that a certain mechanically-minded senior army officer did when shown how!

My boy sometimes fails to obtain a start because he is afraid of over-choking the induction pipe with his finger. I have now taught him a simple way of getting around the difficulty. I can well understand a boy may not be able to judge very well whether the mixture is too rich or too lean to start. My boy now opens the needle-valve half an extra turn. He then drips, with a fountain-pen filler, about three or four drops of *fresh* fuel from the fuel bottle (*which he at once corks up*) into the induction pipe. He increases the compression by a quarter turn. He then starts the motor with a number of swings until the slight excess of fuel clears itself and it fires. As the engine gets going, he returns the needle-valve to its original run position, i.e. half a turn closed, and he slackens off the compression lever the quarter turn he had screwed up to start. He has his perspex tank half full to start and then watches the engine run until the fuel comes down to a red line we have painted on the outside. He then releases the model knowing that there is no timer that can stick and fail.

The engine runs for about half a minute more, and he gets quite a long and exciting flight over a piece of heathland, with a good run after the model to keep him alive and energetic!

Fig. 70. See if the fuel mixture is adjusted for severe climbs before releasing the model.

I always impress on my son that, if he wants to carry on to a ripe old age, with both eyes intact and no scars on his face, it is a sound plan when running-up model engines *not to stand with one's nose right over the top of the cylinder and not to stand at the side of a running engine with one's face in line with the propeller.*

I do not consider this to be over-cautious, because I have seen things fly off during my many years at the game, and after all is said and done, one is always advised not to stand in line with a full-sized revolving aero. propeller! It is just ordinary common sense, like not looking down the business end of a gun although you think the thing is not loaded. There have been many accidents through the ages because these simple precautions have been neglected. I recently saw a new experimental diesel's cylinder-head blow off.

A point that all competition aspirants should attend to is the simple test of holding the model with the engine running for a few moments in a stiff climbing attitude, to see if the mixture control is correctly adjusted for such a position which the plane may adopt during a steep climb. (See Fig. 70.)

REDUCED THRUST FOR TEST FLYING

If the propeller is bolted on back to front, it will reduce the thrust of the engine for the first test flights. When trim is found, the propeller can be put on the correct way and full thrust used.

A FEATURE OF DIFFICULT STARTING

Very often a diesel is difficult to start for the first run of the day, whereas once the engine has run, it will start very easily for flight after flight or run after run during the day, in spite of the fact that the engine becomes absolutely dead cold between runs. There is a very simple reason for this that, if not known, may cause considerable trouble to the novice.

Diesel fuel is sticky stuff and when the ether evaporates from the fuel mixture, there is generally a sticky oily mess left in the little tube that runs up from the bottom of the tank to the *very small* orifice that the needle-valve seats upon. When the new fuel is put in the tank for the first run of the day, and the engine is choked by the finger, the suction from the engine is not sufficiently powerful to clear the blocked passage. The remedy is simple. Before starting up for the first run of the day, screw down the needle-valve and note the exact number of turns it was open for the running position on the last occasion if you do not already know this setting. Then unscrew the needle-valve and take completely out. Put a piece of rubber tube about 6 in. long over the orifice and blow down the tube. This will clear any gummy deposits. Replace the needle-valve, and *open it to the "correct run position."*

Now start up in the normal way and fuel can then be sucked up correctly.

When screwing down a needle-valve, do not be heavy-handed, or the seating and needle may be damaged.

PUTTING THE DIESEL AWAY AFTER FLYING OR BOATING, ETC.

There is no sparking plug, as on a petrol motor, for oil to

drain into, or ignition points to become saturated in oil, that will give trouble when we next take our diesel-engined model out for a day in the air or on the water.

But we should remember to empty out the fuel tank because the ether in the fuel will evaporate and leave a difficult starting fuel for next time. Also a diesel is a dirty motor and throws out diesel oil and lubricating oil over itself and the model. Therefore a quick wash down with a little petrol and a brush and a rag to dry off is not a bad plan, at the end of a day's fun. Also wipe off the fuel from the wings and fuselage of the aeroplane or the paint and varnish of a boat hull or car, because the fuel stains paint and varnish and rots silk or nylon wing coverings.

Fig. 71. Diesels can be mounted inverted.

Make sure you cork that fuel bottle securely up each time you use it to keep the ether content in the bottle! Even then it does no harm to add a dash of ether to the fuel and shake, if the corked bottle has been put away for some time—starting will probably be a little quicker.

INVERTING A DIESEL

Many aeromodellers wish to invert their engines because they like having the cylinder positioned low down, partly for realism, and partly because the weight is low, and sometimes because it suits a neat line of cowling. The inverted motor also permits of a high thrust line.

I am often asked, " Can I invert my diesel ? " The answer is, " Yes." I have inverted most of mine at one time or another. There are two possible snags—neither of which are insuperable difficulties.

The first is on a contra-piston engine. The compression adjusting lever is not so readily to hand, although it can be operated with a little practice. The second snag is our old bugbear, a surplus of liquid fuel more easily gets to the cylinder-head and causes the " over-hard " engine, and therefore sometimes causes extra labour in this matter, as described in Chapter IV, " How to Start Diesels "—But forewarned is forearmed.

A clever scale control line model with inverted engine by Dr. Thomas is shown in Chapter VI. Dr. Thomas finds inversion helps his starting. I have found the same thing on several engines. It is equally possible to mount a diesel on its side as the control line enthusiasts do, calling it a " sidewinder."

THE " FIELD GEAR "

We all realise that one of the great advantages of the diesel is the elimination of a booster accumulator and flight battery. Let us look at Fig. 72 and see just what this advantage does mean —on the left of the photograph you will see the box containing booster battery, spare flight batteries, sparking plugs, petrol can, propellers and tools with booster wiring, that I find necessary to fortify myself with when I go out for an afternoon's flying with petrol-engined models. This includes several different " hot " fuels in bottles for varying compression engines. On the right of the picture, I have laid out the gear which I take when the same model has a diesel engine fitted as its power unit ; is it surprising that the diesel is a popular engine ?

The items that are necessary in the latter event are : 1 fuel

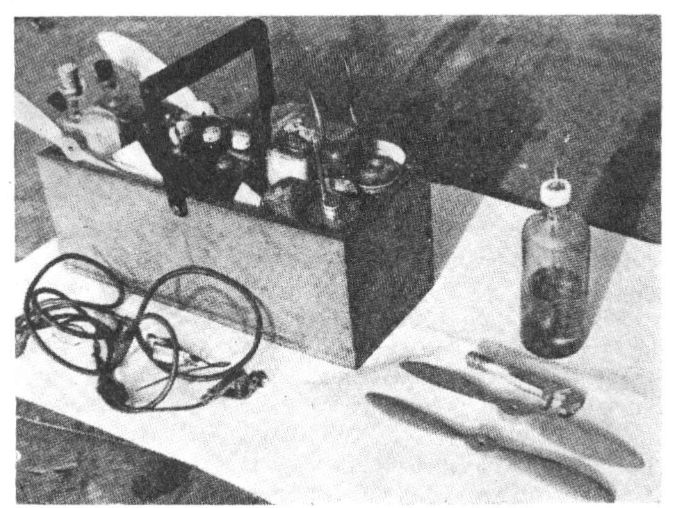

Fig. 72. The field gear carried by the author for a petrol model is seen on the left. The simple gear carried for a diesel is on the right.

bottle ; 1 small spanner to fit propeller retaining nut (possibly a pair of pliers) ; 2 spare propellers, just in case !

CONTROLLING THE DURATION OF RUN OF A DIESEL

When I set up the first officially observed controlled record flight after the 1914-1918 war, I used a weird and wonderful clock device that I had adapted. This clock operated a long arm, which in turn pulled back the throttle lever of the 28-c.c. two-stroke petrol engine that I used. The engine was large and controllable, and was capable of a steady tick over with its 24-in. diameter propeller. This was the first time that a model aeroplane had been controlled to throttle back after a flight of a predetermined time ; in this case 50 seconds.

The extraordinary thing about the whole episode was that it worked well.

In fact, so well that my first record flight was only of 75 seconds, the last 25 seconds being glide with engine ticking over.

The electrical ignition was then cut off by a long arm between the undercarriage legs as the biplane "Kanga" actually touched down. Shortly afterwards I put up a long record flight of over 12½ minutes out of sight. This time I fitted a clock that cut the electrical ignition, because I fitted the first small engine of 15 c.c. to fly, and this engine would not run slowly.

Several of my competitors in those early days relied upon the engine overheating and fading in the air to terminate the duration of their flight! Later on in the history of power-driven model aircraft it very rightly became a rule that an efficient time-switch must be fitted by all S.M.A.E. members, and also before a model could be insured.

In spite of these precautions, people today constantly allow their models to "fly away" because of inefficient timing devices or because of none at all.

It is immensely important for the model movement in general that aeromodellers particularly should control the duration of their models. A model that flies into the blue is a danger to fellow human beings and property. "An Englishman's home is his castle," etc., and nothing infuriates him more than strange objects, aerial or otherwise, arriving uninvited through his windows or into his garden. No man other than an ardent aeromodeller likes to be pursued by a screaming diesel!

The matter must be taken seriously, and not left as an after-thought, as so many modellers do, if restrictive legislation is not to occur. Apart from the other man's feelings, it always seems to me a foolish pastime to build an expensive model and then lose the creation of one's brainstorm on the first day out. I know I do my best to control mine, and in addition, like the wise virgin, I metaphorically trim my lamp by adding my name and address to the fin of the model with a little reminder that there is a *reward* for the finder who returns the model to me, just in case. As a result, the models do not often fly away!

The petrol model aeroplane that has its duration of engine run stopped short through a time-switch cutting off the ignition to the sparking plug suffers from the great disadvantage of often being in a steep climb when the engine cuts with a suddenness that does not allow the nose of the aeroplane to get into a steady gliding position. The resultant series of stalls is often most distressing for the owner to watch. That was why I fitted my first large record machine with a gradual throttling device. Unfortunately, the smaller petrol engines of recent years will not throttle easily. Hence the ignition time-switch. I now design my models to level out under these circumstances, as this is possible to do.

The diesel has no ignition to cut, therefore designers and owners have evolved various devices to cut off the fuel supply or to choke the air intake, or to admit air below the jet to destroy suction. These methods are superior to the ignition switch of the petrol model because, where the fuel is cut off, a little remains in the feed-line and the engine dies down with pops and irregular running, and in the case of the choke, the mixture becomes starved of air and very rich. Again, the process of stopping the engine takes a few seconds during which time the aeroplane's speed drops, and so does the nose into its gliding angle. Perhaps the best method is to close down the fuel supply by a valve in the line.

There is only one real danger that I have found in the case of the choke or the air-leak method. The engine sometimes manages to draw in just sufficient air to run very badly and keep on running. In this case you really have " had it," because the model, if lightly loaded, carries on flying quietly away for an even longer duration due to the lower fuel consumption. I therefore prefer the positive fuel cut-off, or a limited tank capacity. For radio flying see Fig. 59A.

Besides having a timing device and my name on the fin, *I therefore also see that the fuel tank does not hold too much fuel for a really long flight should the switch fail.* A boy in our local club lost two expensive diesel engines in two days' flying : one cost over £8, and the other over £6 10s. 0d. This hobby of carelessness is more than the average person can afford.

How does the modeller set about fitting up his engine to stop after a predetermined time ?

There are two stages. First, the device on the engine to

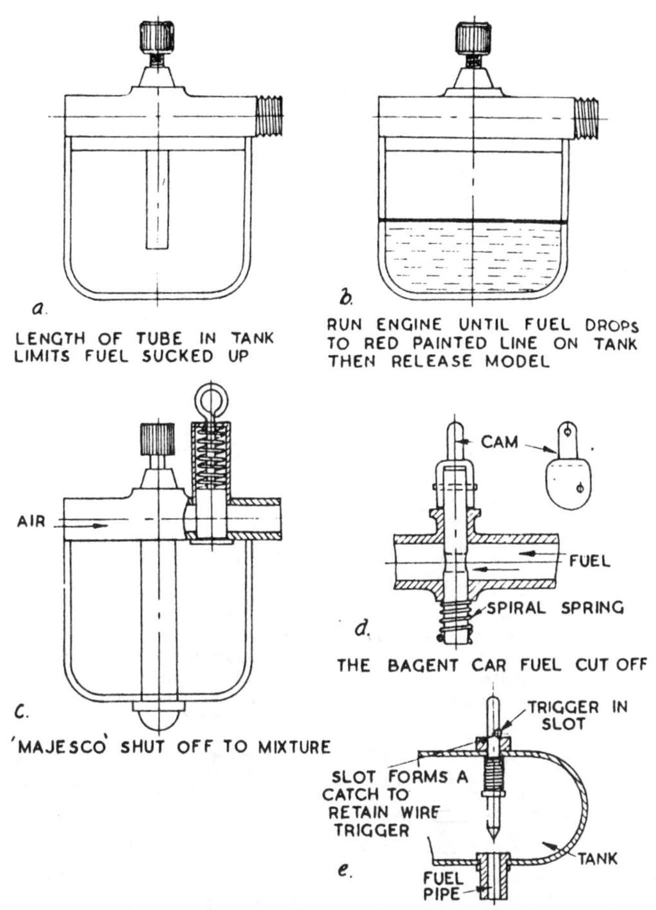

Fig. 73. Methods of stopping a diesel. C, D, E have to be operated by one of the mechanisms shown overleaf.

choke it, cut off its fuel, or destroy the suction on the needle-valve jet, and secondly the mechanism to operate this device. Although not fitted to the early diesels, most modern commercial engines are now being fitted with some sort of stopping device.

That fitted to the German " Eisfeldt," originally pointed the way to various makers. In this method an air-leak hole is opened around or below the needle-valve.

The British " Mills " engine has adopted a somewhat similar method.

In Fig. 73 I have sketched a number of simple devices I have seen or tried. A glance at the sketches will be more satisfactory

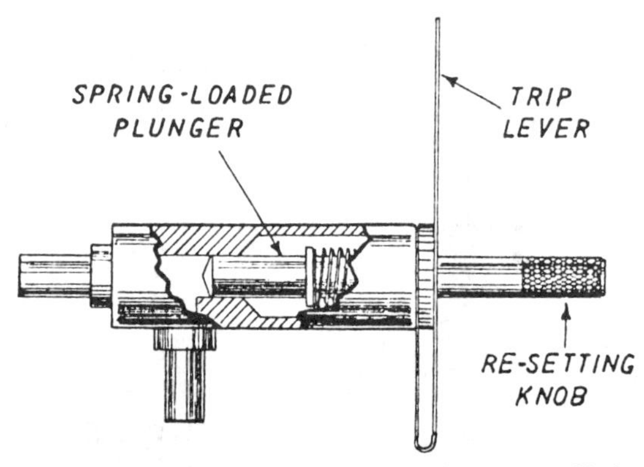

Fig. 73A. The "F.G." fuel shut-off valve cuts off the fuel to the needle-valve and is certain in action, although it makes the engine stutter to a stop which gives the model aeroplane time to get its nose into a gliding position.

than a long-winded description. Very tiny diesels are best controlled by the baby tank method shown in A and B in order to save weight.

One of the snags of some of these devices is the power required to operate them. A particularly clever and simple idea is that of the " Micron," but it is difficult to find a time-switch that has the strength to operate it. In the case of the " Mills," I have seen the return spring removed in order to allow a light-weight " Snip " timer to operate it. The " Majesco " engines are fitted with a very light spring which can easily be tripped by a small airdraulic timer. The " B.M.P." engine has a delightfully easily operated little flap over the induction pipe. This has a

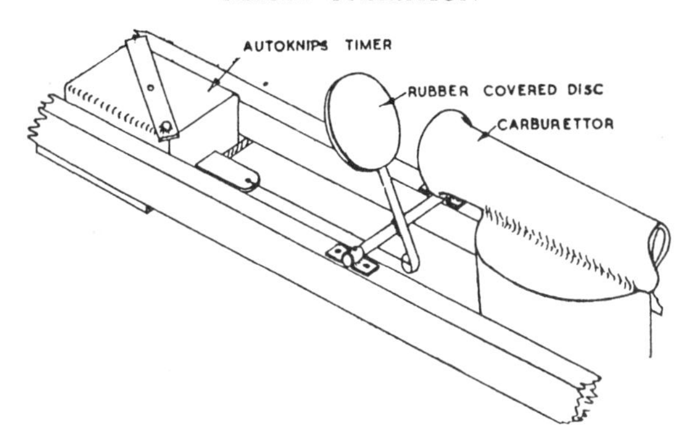

Fig. 74. An " Autoknips " timer rigged up to close carburettor intake. The E.D. commercial clockwork timer can now be obtained to do the above work.

rubber pad on its face to prevent any air leaking past the shutter flap. One of the best and most simple devices I have seen is Mr. Baigent's fuel cut-off as fitted to his 4-c.c. diesel car. This can easily be made up by the mechanically proficient, and is quite foolproof as well as having a light operation by any small time-switch. The new " Frog " diesel cut off constructed with the filler cap is simple and effective. It operates on somewhat similar lines to that shown in Fig. 73 (*e*) or in Fig. 73A.

The operation of these time control devices can be done in various ways. Dr. Forster has devised a simple air vane or little metal propeller driven by the slipstream of his models that operates his time-switch, through a helical screw that sets off a train of gears. Somewhat similar devices are used by the French. There are various airdraulic time-switches available on the British market, such as the light-weight " Snip " timer, the " Gremlin " and the " Majesco " time-switch. These all work on the controlled air-leak principle. They weigh very little and are reasonably cheap, and they are accurate to within a few seconds, but cannot be called dead accurate for competition work in cases where a flight to a certain second is nominated.

An excellent air switch called the " Elmic " has recently been

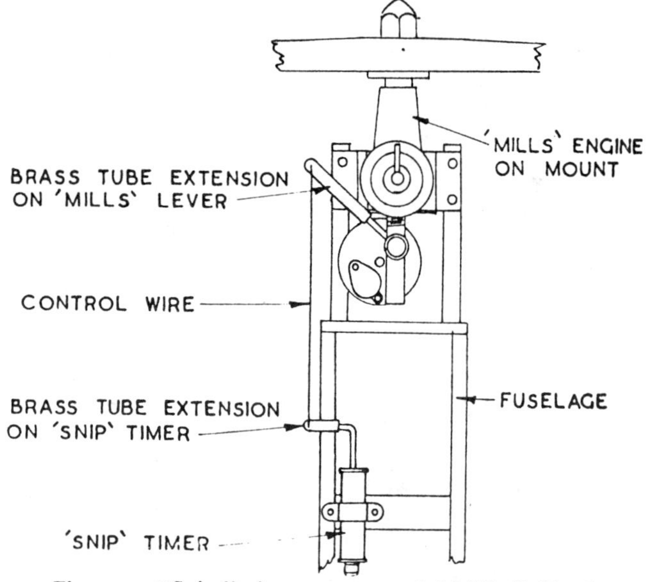

Fig. 75. "Snip" timer to control "Mills" diesel.

put on the market. It has a high degree of accuracy and an excellent device to hold it open until it is released.

The E.D. clockwork time-switch specially made for model work, and on the British market, solves many difficulties where accurate timing is required for competitions. Every second counts! Refer back to Fig. 59A.

The timing devices shown are, of course, also suitable for boat work.

Very small plastic tanks like baby fountain pen fillers are used to give approximately 30 seconds' motor run. I recently saw that an owner had taken three turns of neoprene tubing around his "Frog" diesel tank. A larger tank was used for starting. Just before release of the model, this was withdrawn and the engine run on the tube full of fuel. What could be more simple and a greater insurance against a flyaway.

ENGINE COWLINGS

Most people like cowlings. They always begin by building a cowling for their first model, anyway! Later, many discard them as a nuisance on flying model aeroplanes and working boats. Cars are somewhat different.

The diesel engine runs much cooler than its petrol brother

Fig. 76. Metal spinner and cowlings can be made from aluminium spun or beaten into shape. The cowling on this lovely flying scale "Lysander" adds much to its realism. The motor is a 2.4 c.c. E.D. diesel.

and it has no sparking plug to short to a metal cowling. It is therefore a far more simple matter to cowl a diesel efficiently.

Most sheet-metal firms will spin up a cowling on application for a few shillings. A suitable cowl to suit the nose of one's model can also be beaten out by a panel beater for quite a reasonable sum. I use both these forms of cowling, but must admit I often fly without any cowling!

Cowlings certainly improve the appearance of models enormously. Varied spinners are now on the market.

Rather a beautiful example of cowling a diesel engine on a scale model can be seen in Chapter VI, which shows Dr. Thomas's

" Typhoon " well cowled. This model is a " control-liner " and always flies with cowl in position, which is made from carved balsa wood. A simple metal cowl can be seen fitted to my kit model " Meteorite " in Fig. 88, Chapter VI. Also see Figs. 76, 76A and 76B in this chapter.

TROUBLE FINDING

The diesel is so simple, with its lack of ignition gear, that it is not difficult to pin down faults if they are followed through in an orderly manner. Below is given a summary of the most likely troubles that may be met with :—

(1) *Poor Compression.* This can be heard escaping if the propeller is held halfway against maximum compression, or if bubbles of oil show through exhaust port, escaping between cylinder and piston. I have found this on one or two commercially-made diesels, but it is not usual, as manufacturers with a reputation know how to fit a piston to the cylinder.

This trouble is most usual on amateur-made diesels. Even experts quite used to full-size production do not at first realise the very accurate fit between piston and cylinder that is required for a model engine—cylinder bores must be absolutely round and parallel.

The only cure for this trouble is fitting a new piston to a round cylinder, the piston to be made to finer limits.

(2) *Gumming Oil.* This gums up the needle-valve when the engine is put away, when the ether evaporates. It can also make the motor itself very gummy and difficult to turn.

Cure : Take needle-valve out and clean. Blow through suction fuel tube. It is seldom necessary to clean out the engine and never necessary in the case of fuels I have recommended.

Some of the curious gummy mixtures advocated, particularly on the Continent, cause a lot of trouble in this respect. Dirt in the fuel will also cause a blockage of the needle-valve. Filter all fuel.

(3) *Incorrect fuel mixture of constituents.* Leaving the fuel bottle uncorked will allow the ether to evaporate and so leaving incorrect proportions of constituents. Careless mixing of proportions is another frequent cause. The cure is simple. Be careful!

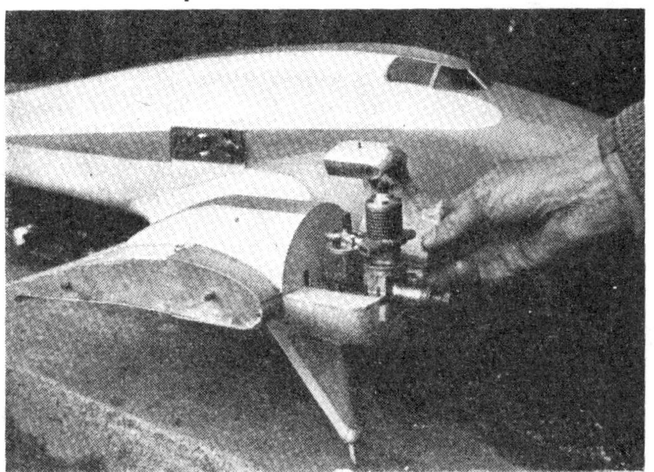

Fig. 76A. A simple carved cowling from balsa wood. Note the hollowed out top cover kept in position by locating dowel pegs and dress snaps, and hinged one side.

(4) *Damage by Acid.* Certain ethers have acid in their make-up. This damages the fuel containers and the engine. Be careful to empty out tanks after use, and blow out as much fuel as possible by rotating the engine a dozen times after use and before putting away. Only buy "anaesthetic ether" or pure "ether meth," as used for anaesthetic work. Neither of these ethers, obtainable from a chemist, have acids in them that will damage an engine.

(5) *Needle-valve.* This may become unsoldered from the thumbscrew top, if made in that way. The operator then turns—what he thinks is the needle-valve—to open or shut, when in fact the needle does not move. This can

be a very puzzling fault and I have known it happen on several occasions.

The cure is to sweat the needle to the brass or other metal thumbscrew. A bent or broken needle can cause trouble. This should not take place if the owner

(*a*) does not screw down too tightly;

(*b*) is careful when taking out the needle-valve.

(6) *Air Leaks.* Due to damaged washers and facings through careless stripping and assembling of the engine. The cure is to renew washers, etc., and not to strip and assemble the engine unless you know how! And leaks can, of course, occur down a worn crankshaft bearing as well as a worn piston.

(7) *Leaky contra-piston.* Amateur constructors do not always realise that the contra-piston must be a perfect fit, on the tight side, or leaks of compression occur *via* the contra-piston and the screwed thread of the adjusting lever.

Sometimes the contra-piston is fitted too tight, with the result that the compression will not force it back to the adjusting lever. One can tell this by finding the lever is loose and slack after it has been turned back and the engine turned several times over compression.

One cure I carry out in such cases is to put some oil in the cylinder and turn *carefully* against compression. The oil being incompressible will force the contra-piston up. *But be careful or you may damage the engine if the contra-piston is really seized up, and you over-force the engine.*

(8) *Excessive vibration when running.* The cause is usually an unbalanced propeller or flywheel, or an insufficiently rigid engine mounting.

(9) *Engine will only fire gently.* On a number of engines there is a device to stop the engine, in some form of cut-off to the inlet pipe. Often this has half closed and is choking the mixture. This is a more frequent happening to the novice than perhaps the reader will believe.

I have seen a very illustrious "petroleer" caught this way when I introduced him to his first diesel!

(10) *Engine becomes stiff.* Use of incorrect grade of oil. A very frequent happening, because people think model engines like very thin oil, or because they want to economise. Lack of adequate lubrication is the cause.

Fig. 76B. A commercially obtained spinner, with hollowed out balsa cowl to blend into spinner and monocoque fuselage is quickly made. Kept in position by dress snaps or quick drying cement which is easily parted and renewed as required. The model is the author's "Satellite" kit monocoque high wing.

(11) *The "Hard Engine."* I have very fully explained this phenomenon earlier in this book, *never force a "hard engine,"* i.e., an engine that is almost impossible to turn over compression due to too much liquid mixture in the cylinder-head.

(12) *Incorrect diameter and pitch propeller.* An incorrect propeller utterly ruins the performance of a model diesel. See my notes elsewhere on this subject.

For full details of propellers, and how to make them,

suitable for various types of flying models, these can be obtained from my book, *Petrol-Engined Model Aircraft*, published by Percival Marshall & Co. Ltd.

(13) *Engine slows up after running.* This may be due to lack of adequate lubrication, but in the case of a diesel is more likely to be due to too high compression after engine becomes warm. Try slacking back the contra-piston lever slightly.

(14) *Engine bounces to and fro when starting.* Compression is high; release slightly.

(15) *A puzzling trouble.* The engine starts after choking and then stops after a short burst. It repeats this process and will do nothing else, whatever aids are given. First make sure that the compression ratio is not too high. The fault is almost invariably due to the small hole drilled in the jet tube having become turned away from the air stream in the venturi tube of the " carburettor." There are a number of engines constructed so that the jet tube is rotatable. The carburettor must be taken down and the tube rearranged with the small hole for fuel arranged so that fuel flows evenly through the venturi tube, i.e., the hole must face the rear *and must not be sideways.*

CHAPTER VI

AEROPLANES, BOATS AND CARS AND RADIO CONTROL USING THE DIESEL

THERE are very few people who buy an engine without the intention of fitting it into a working model. The diesel is particularly versatile. It can be used in model aeroplanes from large models right down to 21 in. span models, and less. It is ideal for flying-boats and seaplanes because of the elimination of electrical troubles due to damp. There is none of the troublesome festooning of long booster battery starting leads around the legs in a dinghy. As soon as a start is made the model can be placed on the water, and off she goes.

Having done a considerable amount of diesel flying off and over water, I consider that the diesel has absolutely revolutionised the model waterplane. It has brought this intriguing side of the hobby within the range of many more people. Now that all the old petrol paraphernalia need not be carried, and the waterplane can be built in really small sizes due to the elimination of ignition weight, should a modeller not happen to own a small boat, there is no reason why he should not hire a dinghy on any inland water or protected stretch of sea water. He can then thoroughly enjoy a day's aquatic flying pleasure.

The diesel is also the answer to the power boat enthusiast's prayer. I have thoroughly tested diesels out in model speedboats of the hard chine class and also in the hydroplane class. They are most successful in the hard chine class of boat, and I believe that in the future they will compete in the round the pole hydroplane racing now that the high speed racing diesel has materialised.

The diesel engine is particularly suitable for the hard chine speedboat because it runs cooler inside a hull than the petrol engine, and because water does not attack the electrical ignition gear that previously had to be carried.

Fig. 77. The author's free-flight cabin low wing monoplane for 1 c.c. diesels.

It must be remembered when operating over sea water that the salt causes corrosive damage to engines if it is left in them, should there be an undue splash on landing or in speed-boating over rough water. If possible, the engine should be run before putting it away at the end of the day and then thoroughly oiled externally. If a dive into the sea has occurred by some mischance, the engine must be dismantled at once and thoroughly cleaned out, with due care taken to remake all joints correctly, as I have emphasised earlier in this book, *unless it starts up at once and dries itself out before condensation sets in.* My experience in such

Fig. 77A. Here we see the wing tip slots and a diesel doing their work at a high angle of climb on a low wing monocoque model by the author. The wing span is 4 ft. 4 in. and the diesel a 2 c.c. "E.D. Competition Special."

cases has been that the diesel usually does start up even with sea water still in it! This seldom ever happens in the case of a petrol engine.

In Czechoslovakia a small scooter powered by a little diesel engine has been produced. One would like to suggest that the diesel could very easily be commercially produced as a model, and also a miniature full-sized outboard boat motor.

In order to help individual modellers, whether their inclination is towards aeroplanes, boats or cars, I am devoting this chapter to a survey of experimental models that I know of or have built myself and tested with diesel engines of varying

Fig. 78. The author's 42 in. span biplane fitted with wing tip slots is seen climbing lustily powered by a 2.4 c.c. "Elfin" diesel. The model is exceptionally stable partly due to the "slabsided" fuselage.

capacity. I make apology for the large proportion of my own models that are included. This is partly due to the fact that I naturally have an intimate knowledge of the performance and constructional data of my own models which therefore provide useful comparisons. It is hoped that the release of these details will help many fellow modellists to get going with diesel engines.

All the models shown are stable and successful operating models, and I am giving general details and measurements so that they may form a basis of experience for any of my readers who wish to design their own craft, but are not quite sure of the capabilities of different sizes of engine. Because the diesel has more power, *and a different type of power*, and because it has less weight to carry for its increased power, as a result of the elimination of the electrical ignition gear, it is obvious that our ideas of suitable models for any given cubic capacity of

Fig. 79. This tiny control-line model of 17 in. wing span has flown successfully, powered by a "Kemp" 0.2 c.c. diesel. It is a scale model of a "Westland Widgeon."

petrol engine will require revision. This elimination of the ignition gear permits of new conceptions in regard to even smaller models than we have hitherto considered as the limit of practical projects.

The advent of the diesel has therefore opened up new territory for exploration and given the model designer new fields to conquer.

We have mentioned that France has produced one engine of only 0.16 c.c. which permits the powering of a 4-oz. indoor round the pole model. These microscopic engines, however, cannot yet be called practical propositions for the ordinary mortal, because they are too expensive to produce commercially, and are also too difficult to operate.

A SMALL LOW WING MODEL FOR 1 C.C. TO 1.3 C.C. DIESELS

A stable low wing model looks like the real thing in flight, and if properly designed can be very stable in spite of the

Fig. 80. This view of the "Humming Bird" shows a cone mounted "Majesco Mite" diesel of 0.7 c.c. installed. The fuselage is of sheet balsa.

popular belief to the contrary. It is important to keep the thrust line low to prevent excessive torque reaction upsetting the model's stability. The small 48-in. span low wing model shown in Fig. 77 makes a handy size machine. This model is flown by a 1 c.c. E.D. "Bee" diesel. A low wing model takes off the ground straight after quite a lengthy run, and the "cushioning" effect of the low set wing makes good landings a modeller's dream. They are not noted for a rocketing climb.

Readers of mature age will remember the successful little full-size "Westland Widgeon" light aeroplane of between war years. Dr. Thomas has built a tiny control line model of this craft, which is seen in Fig. 79. It is powered by the midget Kemp 0.2 c.c. diesel, and has the small wing span of 17 in. The length is 12 in., and weight is only 3½ oz. This tiny control liner flies successfully in calm weather on short lines. Naturally there is not enough weight to keep the lines taut through centrifugal force in windy weather.

A well-known full-size aeronautical journal, remarking upon

Fig. 81. Find the baby! A 0.2 c.c. " Kemp " diesel is sitting on the starboard wing of this control-line model. It makes an interesting comparison with the large 5 c.c. American " Drone " diesel which powers the author's model.

the capability of the little 0.2 c.c. Kemp in flying a 2 ft. wing span model, pointed out that, scaling this up to full size in proportion, it is equivalent to flying a " Tudor " with an engine of only 12 c.c. ! This shows the great power for c.c. these baby model engines produce.

Before the birth of the diesel of very small capacity, the lure of the baby free flight model gave model designers many a headache. If the model was made satisfactorily small the weight made it fly too fast to obtain satisfactory stability. The advent of the tiny diesel changed all that overnight. Fig. 80 shows a little 36-in. span high-wing cabin model which I produced for diesels of around 0.8 c.c. such as the " Amco," the " Allbon Dart " and the 0.7 c.c. Mills. This little machine is called the " Humming Bird " and is now available as a plan from the publishers of this book. The model is very stable and the fuselage is made from balsa sheet in spite of its small size, with the result that it is nearly crash-proof. It will fly in really high

Fig. 82. The new "Hivac" valve radio sets encourage small radio models powered by diesel engines. The author designed his 48 in. span "Radio Brumas" for the latest lightweight radio sets. Engine is a 1.8 c.c. "Elfin."

winds and suffer no damage on landing if reasonably trimmed. Sheet construction allied to light weight see to this.

Fig. 81 may interest readers with an enquiring type of mind. The question is find the diesel. On the starboard wing a 0.2 c.c. "Kemp" diesel is seen sitting as a comparison in size with the 5 c.c. American "Drone" diesel installed in the nose of the model.

BABY RADIO MODELS

The year 1950 has seen the British baby radio sets arrive, due to the latest "thyratron" gas filled mini-valve. It is now possible to fly little models by radio in this country. My little "Radio Brumas" seen in Fig. 82, has a wingspan of only 48 in., and was developed from my larger radio aircraft. The diesel makes the best power unit because of its light weight. Larger radio sets are by no means defunct or outmoded. They and the larger model fly with greater steadiness and easier tuning if the "raised current" system is used, as described at the end of this chapter. Fig. 83 shows one of my medium size radio models powered by the high powered 3.5 c.c. Mark IV

E.D. diesel, with three valve E.D. radio receiver. The wing-span is 78 in.

CONTROL LINE FACTORS

Although a higher wing loading than for free flight is desirable, it is a great mistake to think, as a number of people do, that an excessively high wing loading is desirable for stunt models of the control line variety.

The Americans proved that a lightly loaded model makes for better stunting and indeed speed flying. This is because a lighter model is moved from its path more easily when the controls are altered. The heavy model " sticks in the groove." When one says lightly loaded this should read as reasonably light loading. Therefore reduction of weight is important. Another factor proved in the great American control line boom, and often lost sight of over here, is that a streamline symmetrical section makes for smoother control flying, greater speed, better stunt performance, and far better landings than the " flat plate " thin wing section so often used with wings made of solid balsa.

I have a garden circuit where I can fly whenever the spirit moves, or when a friend with control line aspirations comes

Fig. 83. The larger diesels suit medium sized radio models. The author's 78 in. span radio model is powered by the high performance E.D. 3.5 c.c. Mark IV diesel and controlled by an E.D. radio, modulator three-valve set.

Fig. 84. The "Bowden Bullet" control line kit model is suitable for medium-size engines, such as the "Frog 100" or "180" diesels, the "Elfin" 2.4 c.c. diesel and "Amco" 3.5 c.c. motors.

visiting. As a result I have done a great deal of control line flying and experiment with varying wing sections, propellers, engines and so on. I have found the American ideas outlined above completely revolutionise control line flying. Should a reader have previously flown only a thin wing of the flat plate type, I urge him to try a built-up symmetrical wing sectioned model.

The model shown in Fig. 84 will give an impression of what I mean. It is called the "Bowden Bullet" and has a span of 23 in. with built-up sheet covered wing. It is monocoque and can be built from solid laminated balsa hollowed out, or it may be planked. A model has been made on both principles and both weigh approximately the same, the planked version being a little lighter and, I think, easier to build.

Mr. Phil Smith has designed a delightful series of control line models of the near scale type for the well-known firm of kit manufacturers, Veron. He employs an interconnected wing trailing edge flap with a smaller elevator than is usual, which

Fig. 84A. Mr. Phil Smith's "Midget Mustang" won the British International Team Race of 1950, powered by a 3.5 c.c. "Amco" diesel, using a 9 in. by 6 in. pitch propeller.

Fig. 85. Another Smith control line model is the Focke Wulf 190, A3, powered by a Mark IV, 3.5 c.c. diesel. Note the trailing edge flaps on the mainplane which go down when the elevator goes up, to assist quick stunt work.

gives smooth but very quick stunt movements. As the elevator goes up the forward flap goes down. Fig. 84A shows one of his models for team racing. This model, the "Midget Mustang," won the first British International Team Race, held at Brighton, in 1950. It is now sold in kit form.

The pure speed enthusiast should remember to get his control line model to fly in an absolutely flat path, with its streamlines in line with that path. A very high pitch propeller will be required, whereas stunt work requires a pitch in the middle ranges.

MODELS SUITABLE FOR DIESEL ENGINES OF 1 TO 1½ C.C.

A Baby Flying-Boat

This is a very popular size of motor because these engines are not quite so touchy for the novice to operate, yet they are cheap to buy and the model can be reasonably small and portable.

My little flying-boat of 36 in. span can be seen in Fig. 86, powered by a 1-c.c. "Frog" diesel, or 1.49 c.c. "Elfin" diesel. The model is called the "Wee-Sea-Bee"; plans are now commercially available.

Small models like these are a joy to operate because they are so handy when starting operations are in progress from a full-sized dinghy, or at the pond-side. The hull of this flying-boat is a "monocoque" and is made entirely of $\frac{1}{16}$-in. sheet balsa. It is fitted with sponsons for lateral stability on the water. The all-up weight with a "Frog" diesel fitted is only just 16 oz., and the model planes very quickly and easily when taking off. I noticed one of these little boats had been converted to a baby amphibian with detachable wheels by a competitor at the 1948 Bowden International Trophy.

It should, however, be remembered that a baby flying-boat is certain to require very delicate trimming of thrust line, etc., whereas a larger boat is far less touchy. I have had a number of exciting and very high flights from this little boat. Not long ago she flew from in front of the Parkstone Yacht Club over Poole Water, and the wind changed after take-off. The model headed with a full tank of the "Frog 100" diesel towards the

Fig. 86. The author's baby 36 in. span diesel-engined flying boat, " Wee-Sea-Bee."

port of " Poole," climbing fast and, much to my concern, for I could visualise the model landing in various awkward places in the main streets of Poole, perhaps to the fury of the towns-people of this ancient port. I devoutly hoped their spirit of seafaring adventure for which they are famous, would stand me in good stead. However, the thundery wind changed again and I saw with great relief the speck in the sky turn towards the yacht club. After a long slow glide in a huge sweep the tiny flying-boat eventually just skimmed over the boat sheds and actually made a perfect landing on the water. The story has its lessons, and one is that a very small boat must have a fair amount of offset to the thrust line in order to get the model to plane over the water straight. If there is a turn, one sponson will dip and the little craft will slew round. This straight run, with propeller torque more or less cancelled out, gives a rather straight flight under power which may cause exciting moments like the one I have recounted. A considerable number of people have made this model of the " Wee-Sea-Bee " because of its small size, in spite of my warnings that a baby flying-boat is a far more touchy problem than a larger one. Some have told me of their " wonderful " flights. Others have wondered why they cannot get their model off the water. One man gave his replica upthrust to get it off. I have proved that upthrust will keep a flying-boat firmly stuck on the water. I made some tests on my larger and quite foolproof flying-boat. A little down thrust, contrary to general expectations, gets the hull planing, for it builds up water pressure below the forward step. This is due to the thrust line having to be comparatively high on a single-engined model flying-boat. Disbelievers should try it and see. The same thing applies to jets. It is useless to point the reaction line upwards. It must be pointed downwards, i.e. towards the nose.

Living, as I do, near the water with boats available, I have spent a great deal of time on flying-boats and seaplanes, and have now finished on a book on the subject.

A good number of years ago a boat of mine was the first powered model flying-boat to take off the water in the world's history. It set up an officially observed record at the same time. I still adhere to the usual form of steps that I used in those early days. There are three steps. The nose to the first step is well V'eed to break up shock on landing. This first step is well ahead of the main step which is situated at the C.G. position. The forward step prevents any tendency to nose in during the take-off and if there is any popple on the water, it helps to bounce the nose up and so accelerate planing of the hull. The main step I usually keep flat or only slightly V'eed. The rear step I keep flat on baby flying-boats and V'eed on larger ones. Some few years ago I raised my own record officially with a modern streamlined flying-boat powered by a 3.5-c.c. diesel engine. See Fig. 103. This has since been successfully attacked by a " Mills " diesel engine boat built by Mr. Gregory. There are now new rules for records which limit the engine run and have spoilt the fun!

If the reader will turn a few pages to Fig. 94, he will see a slightly larger 48 in. span flying-boat which I designed for engines like the 2.4 c.c. " Elfin " or 3.5 c.c. " Amco." This size

Fig. 87. A small monocoque free flight model for diesel engines of between 1 c.c. to 2 c.c. The " Bowden Satellite " 48 in. span.

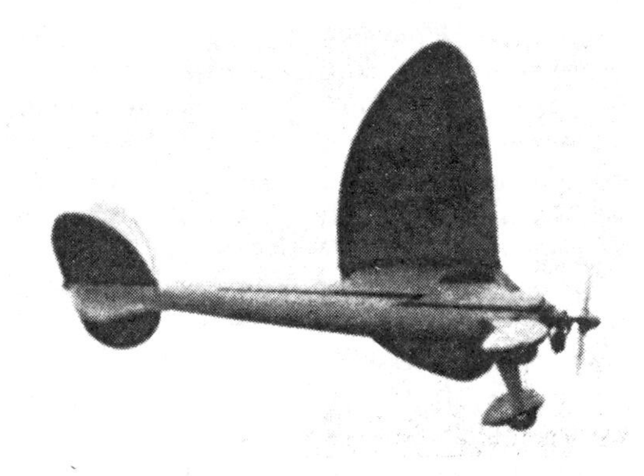

Fig. 87A. The " Bowden Satellite " has been caught by the camera in flight powered by a " Frog 100 " diesel.

boat is very handy to transport and it will doubtless eventually find itself under radio control with one of the new baby radio receivers installed in that capacious cabin. The sponsons are built like a wing and held to a cut away portion of the hull's bottom by rubber bands like a low wing. This prevents damage in the event of a dive in, due to bad trim. Water is very hard stuff and can wreck a rigid structure !

A Small Monocoque Model Landplane

Figs. 87 and 88 show the size of model aeroplanes that I have found most suitable for stable slow flight for the class of engine under review.

I always feel very much against fast-flying models for new-comers to the game of free flight, because fast-flying models suffer much damage and give little satisfaction in the hands of the novice. The problem in design of the smaller type model is to obtain a slowish stable model that is also robustly built. Not an easy combination to obtain, and incidentally, too infrequently attained.

My answer is in the two models seen in Figs. 87 and 88. I have built innumerable models in this class, and have arrived at these two models as a result of what I consider to be the ideal combination to date. Life is always progressing, however, and I may not think so later on ! The first model is of the streamlined monocoque type for those who like them. The wing span is 48 in. This model has been kitted and is probably the first monocoque fuselaged model to be released to the public in this form. The oval fuselage is built by planking on a simple method that presents no difficulty to the average individual. The model really does fly with first-class stability free flight, and has also been fitted with a baby radio receiver for radio control. The side areas and adequate keel surface have been correctly worked out to insure stability. Oval fuselage models are not always renowned for their stability. The rectangular " slab-sider " is easier to get right in this respect, hence their preponderance amongst the designers' choice.

Slow flight is obtainable using a 1 c.c. " Frog " or " E.D.

Bee" diesel, whilst an exciting performance results from fitting a 1.8 c.c. or a 2.4 c.c. "Elfin" diesel. The glide is amazingly flat due to the model's aerodynamic cleanliness.

Fig. 88. The "Bowden Meteorite" is a small $45\frac{1}{2}$ in. span plane now commercially obtainable in kit form. It has a great reputation for stable flight. The wing-tip slots can be seen painted in a dark colour. It is suitable for the latest baby radio sets.

A Beginner's Small Model Aeroplane, the "Meteorite"

I produced the second model to suit any of the popular class of 1 to $1\frac{1}{2}$-c.c. diesels, such as the "Frog," the "Mills," and the "Elfin" and also for the "Frog 160" glow-plug engine or the American babies like the "Arden." It is foolproof to fly and to build, having exceptional stability features and great robustness for its small size, and yet it possesses a glide that fills its owner with satisfaction. Many have now been built as the model is available on the market in kit form, and can have a 2 c.c. diesel fitted if desired, in which case a slightly larger wing is fitted.

The fuselage is built up on $\frac{1}{16}$-in. sheet balsa and erected by means of a simple jigging method that ensures accuracy of the vital angles of incidence of wing and tail, a feature that novices so often get muddled over and yet is so vital in a small model. The sheet balsa is covered with paper and is almost indes-

Fig. 88A. A modern baby radio receiver with its batteries fits very well into the "Meteorite."

tructible. I personally dislike the more usual balsa longeron machine covered with paper only. Such a machine often spends most of its time being repaired. The type of construction I advocate weighs only an ounce or two more in this size of model, which is more than allowed for in the wing area provided, and the model then lasts for ever. Due to the elliptical wing, it will be noted that the model has no greater wing span than the normal model for this size of engine, and yet it has a greater area and therefore slower and less dangerous flight.

Flying, and not repair work, is the thing that should be aimed at, especially for the novice, and for the competition fan, who must have reliability if he is to win events.

The wing span is $45\frac{1}{2}$ in., and the central chord $10\frac{1}{2}$ in. Wing-tip slots of a very simple design are fitted for super stability. The fuselage is 34 in. long and the tailplane has a span of 19 in. The flying capabilities of this model are definitely what the doctor ordered for the novice and the competition man who pins his faith to the small model. It has recently been fitted with a baby " E.D." radio set.

Scale Control line Models

The diesel is particularly suited to control line flying because the engine keeps cool when well cowled, and because cowling does not short an H.T. lead, as it is inclined to do in the case of a petrol engine. The H.T. lead and all the other electrical

Fig. 89. This is a very nice-looking but practical control line flying model, by Dr. Thomas. The undercarriage retracts and bombs are dropped by a third control line.

leads are always a difficulty to dispose of neatly in the case of a cowled petrol engine if they are also to be accessible. The diesel requires none of these things.

Dr. Thomas is a keen British exponent of the control line model and his scale models not only look well, but they are also practical flying models. Fig. 89 shows a very nice scale " Typhoon " fitted with a 2.2 c.c. " Majesco " diesel, the inverted cylinder-head of which can just be seen lurking inside the cowling below the three-bladed propeller. The contra-piston is " get-at-able " through a small hole in the bottom of the cowling. This model is 35 in. wing span and has a weight of 30 oz.

I should perhaps explain to the uninitiated that wing loading can be comparatively high for scale control-line models because a spot of speed is considered permissible and safe even for non-racing control-liners. The wing loading of this model is 20 oz. to the square foot, which, of course, would be far too high for a stable free flight power-driven model, which should have a wing loading ranging between 8 oz. to 16 oz. per square foot. Dr. Thomas considers that the wing loading for his " Typhoon " seems to be the optimum for control line models other than racing models. He made the three-bladed 10-in. diameter airscrew from sycamore, and obtains a static thrust of $1\frac{3}{4}$ lb. This is interesting because the reader will remember that I obtained $1\frac{1}{2}$ lb. static thrust from the 2-c.c. " Majesco " with a two-bladed propeller. It therefore appears that there is really nothing to choose between a two-bladed and a three-bladed propeller, except that the three-blader permits the advantage of a shorter undercarriage. As explained elsewhere, I use wide shorter blades for flying-boats for the same reason.

Dr. Thomas's model has a ratchet selector mechanism in the

Fig. 90. The author's simple low-wing free-flight model is here seen flying with a 2.5 c.c. diesel.

centre section, and a third slack control line is used to work the gear, which on the first pull retracts the undercarriage, the second releases the two 500-lb. model bombs, and the third pull lowers the undercarriage, whilst a fourth pull cuts off the engine. Nice work! And it works!

The fuel filler cap and needle-valve are both extended to the top of the engine cowling.

Inverted engine enthusiasts will be pleased to note Dr. Thomas's final remarks, which substantiate my earlier conclusions *re* inversion of diesel engines. He says, "I am very pleased with the engine. It starts very much more easily in the inverted position, and often starts first swing even from cold." That is because the fuel gets to the cylinder head quickly.

Low-wing Diesel Model Aircraft

Low-wing design for power models has always intrigued me immensely: a well-designed low-wing model, provided the operator is experienced, flies with just as good stability as the popular high-wing model. A medium sized low-wing model of mine is seen flying in Fig. 90, during wintry weather. This model was built specially for diesel engines of 2 to 3 c.c.

The model is now powered by an "Elfin" or "E.D." diesel, which in this case can be seen mounted in the upright

Fig. 91. The author's 38 in. long planing 2 c.c. diesel power boat is seen sitting peacefully on the water with its diesel brother flying-boat.

Fig. 92. The "Flying Fish" is here seen planing well powered by a 3.5 c.c. diesel, in wintry weather.

position. The model is fitted with wing-tip slots (as described in *Petrol-Engined Model Aircraft*); these slots are simple and they give extraordinary stability. The wing span is 4 ft. 8½ in., with an elliptical wing having a central chord of 11 in. The fuselage is 45 in. long and the tailplane is elliptical, span 25 in., chord 7¼ in. The fuselage is of $\frac{1}{16}$-in. sheet balsa and has a turtle top decking with slab sides.

The Diesel Speedboat

The diesel engine is the most trouble-free power unit imaginable for boats. For years I have fitted my speedboats and hydroplanes with petrol engines. Now, I have gone over to diesels, and I doubt whether I shall use a petrol engine again in this form of model, except for some special purpose like hydroplane racing or boats in the larger class.

The most useful of my power boats is the "Flying Fish," shown in Figs. 91 and 92. It is 38 in. long, with a maximum beam of 9½ in., and is of the hard chine, Vee bottom planing-hull type. The 2 c.c. diesel fitted, planes this boat nearly as fast as most of the 4.5 c.c. petrol engines I have also used in the hull.

For an exciting maximum performance the 3.5 c.c. " E.D." diesel or a 5 c.c. " Eta," or a glow-plug motor of up to 5 c.c. makes the best power unit. For normal pond work the 2 c.c. diesel gives an ample speed with good planing.

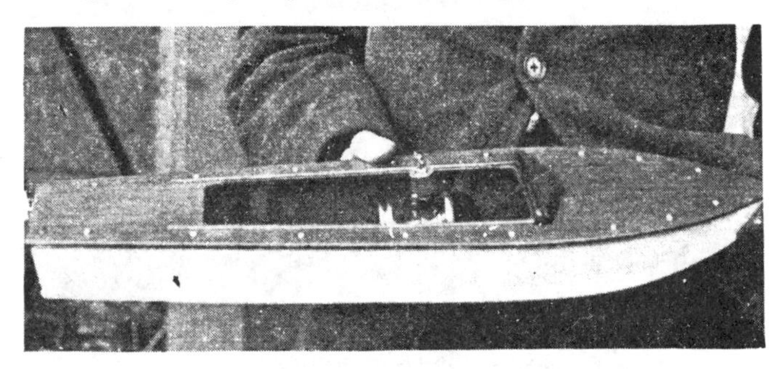

Fig. 93. A 0.7 c.c. midget " Sea Swallow " speedboat hull 19 in. long that planes fast when fitted with a diesel. It was designed by the author. A 1 c.c. diesel gives a racing performance.

The boat is constructed mainly from three-ply, and owing to the demand by people who have seen this rather unusual and yet practical type of model in action, the hull is now obtainable commercially in built-up form or plan form. It is surprising what a very considerable interest has grown up amongst the general public for I.C. power-driven boats during recent years. I have built myself larger and smaller speedboat models of the planing cruiser type, but my favourite has become the boat I have just described, because of its convenient size to carry and operate on pond or sea. For those whose interest lies in a large sea water planing hull capable of dealing with rough weather conditions, Figs. 108 and 109 show the " Swordfish " hull. I now have this hull fitted with " E.D." radio control and a long range fuel tank.

Some people are not quite certain in their minds what a

Fig. 94. The author's 48 in. span cabin flying boat designed for 2.4 c.c. diesels such as the "Elfin." Suitable for radio control with a baby receiver in the cabin.

planing "Vee" bottomed hull means. They visualise a stepped racing hydroplane. This is not, of course, correct. The "Vee" bottomed hull is similar to the familiar and well-known air-sea rescue launches of the war, and the normal pleasure speedboat of peace time, and it has a progressively "V'eed" bottom that becomes nearly flat at the stern. This type of hull automatically and correctly banks on turns and does not skid outwards as in the case of a stepped hydroplane. The baby 19 in. long "Sea Swallow" is shown in Fig. 93. This hull planes well when powered by a 0.8 c.c. "Amco" or "E.D. Bee" diesel. A "Mills 0.75" diesel has recently been fitted into this little hull. These motors give high speed and employ a small propeller.

MULTI-ENGINED MODEL AEROPLANES

The diesel's simplicity of equipment, and as a result light weight, makes the type very suitable for multi-engined enthusiasts. Fig. 95 shows Mr. Crabbe's unusual three-engined model,

Fig. 95. A tri-motor control-line model by Mr. Crabbe. Three " Mills " diesels are installed.

which I have seen him fly control line. It looks impressive in the air, except that so many naked and uncowled engines rather spoil the effect. Small balsa fins had to be located just aft of the outboard motors so that when one cut before the other the model resisted any tendency to turn inwards. One of these fins can be seen on the port wing. The engines are three Mills diesels. It is quite an exciting moment when all three are started and the model takes off. It is a wonder that more modellers do not build multi-motored control line models, for there is little danger when one motor cuts, as there would be in the case of a free-flight multi-motor model. Colonel Taplin has solved the free-flight problem in an ingeniously simple way. He connects his two " E.D." diesels together with a cross-shaft and gearing so that both engines must run at the same speed. He proved the practicability of his method by flying this multi-motored model in the " Bowden International Power Trophy," 1948. A photograph of the connected engines is given in Fig. 96. This model has since been flown under radio control and must be the first of its type to fly in this way successfully. The first flight was one of 8 minutes.

The Pylon Type Model

In America and on the Continent the most popular type of power-driven model aeroplane seems to be that with its wing mounted high on a pylon. This is largely due to the competition rules in those countries, which favour a corkscrew climb on a limited motor run. It always seems to me to be the least interesting type of power model, partly because it is so very far away from full-sized machines in appearance, and further because it has a very unrealistic type of flight. Even the Americans, who introduced the type, are now making critical remarks about its unrealistic flight appearance, in their model

Fig. 96. Colonel Taplin's two-motored model has the two E.D. engines connected together by cross-shaft and gearing which ensures constant speed and thrust from both propellers for this radio-controlled machine.

Fig. 97. An American type pylon diesel model designed by the author, with monocoque fuselage.

journals. Nevertheless, there are a number of people who build this type and I must therefore be careful what I say ! The type has found very considerable support in Britain recently, and a number of manufacturers have put kit models on the market. A glance through the advertisements in model aeronautical journals will supply the needs of modellers interested in the type.

Fig. 97 gives a picture of a "pylon job" fitted with a streamlined balsa-planked monocoque fuselage, and with its diesel engine inverted. The dimensions of this model, which has the usual pylon performance, are wing span 4 ft. 6 in., central chord 11 in.

Flying Wing Tailless Type is very suitable for Diesels

The tailless model has very little drag for an engine to contend with, therefore quite a small power unit will fly this type of model. That seen flying in Fig. 98 is an 8 ft. 10 in. span model which is fitted with a 3.5 c.c. diesel.

Tailless machines are very critical as to weight distribution. This is made simple by using a diesel, because the weight of the

Fig. 98. A large flying wing powered by a small diesel. This model, which was designed by the author, has a wing span of 8 ft. 10 in.

engine is concentrated, and this weight can be located exactly where required. The machine seen in flight has a pusher propeller fitted, in order to get the weight back. Owing to the reduced drag of the tailless model, it is also surprising how fast they fly even when given a very low-wing loading.

The Biplane

Biplanes are still popular in the model world in spite of their having become defunct for full-sized work. They certainly look intriguing in the air.

This type is the exact opposite to the tailless model with regard to drag. Biplanes are truly said to have a "built-in headwind." The small model shown in Fig. 99 is flown by a 2.4 c.c. "Elfin." This model is unshakably stable due to excellent keel surface. The glide is slow and landing excellent, due to the wingtip slots.

The top wing span of this model is 43 in., and it is built with a sheet balsa fuselage.

The Diesel Float-Plane

The 2.5 c.c. class is most useful for the medium-size flying-boat, the hydroplane and the float-plane. I have tried out a

Fig. 99. A small Bowden biplane that flies with rock-like stability powered by a 2.4 c.c. "Elfin." Note wing-tip slots on both top and bottom wing, also very large tail.

Fig. 100. The author's 2.4 c.c. E.D. diesel powered monocoque twin-float plane flies with remarkable stability. Span 48 in.

hydroplane and flying-boat for this size of engine with success. A small float-plane may interest readers, and is seen in Fig. 100. The wing span is 48 in. with central chord of 11 in. The twin floats are planked monocoque and have an extra forward step to assist take-off. I find that the easiest practical method of obtaining a take-off under limited horse-power is to fit floats of a generous buoyancy and large planing surface. This means that the floats do not sink low into the water when at rest, which would necessitate great power to make them rise on to the surface. It is on the same principle that a light wing loading permits flying on low horse-power. A large planing area also makes for easy landings. It is best to forget " wetted surface " in model work !

The diesel engine has an immense advantage for float-plane work because the total weight of the model tends to mount up, due to two or three floats and their strut gear. The diesel's low overall weight therefore prevents too high a wing and float loading.

The engine is a " Mark III " E.D. diesel of 2.4 c.c., and this model looks particularly pleasant in the air, due to its long twin floats.

AERO MODELS AND BOATS FOR 3.5 C.C. TO 5 C.C.

Large model aeroplanes are justifiably noted for their excellent steady flying characteristics, and they are imposing in the air. They require longer to build and are naturally more expensive to produce. Although they fly with great steadiness, the glide must be good, otherwise the weight of the model will cause considerable damage if bad landings are made. Water is a very hard medium if struck at high speed at the wrong angle by sponsons !

The Large Diesel-engined Flying-Boat

An old flying-boat with planked monocoque hull which was originally powered by a 9-c.c. petrol engine, now that it is stripped of batteries and coil, flies well powered by a 6 c.c. German " Eisfeldt " diesel. Large flying-boats require a considerable excess of power to take-off the water.

Fig. 101. A little 36 in. span float plane of the author's which has been flown by a number of baby engines, such as the 0.8 c.c. "Amco" and the 0.7 c.c. "Mills." Three floats are used in this case.

Fig. 102 shows the model in the air flying over Poole Harbour. The model had circled several times around my dinghy in the calm air and finally it came within close enough range of my camera, which for sharp photographic purposes was not quite correctly focused ; the light was also failing, hence the somewhat blurred outline. Nevertheless, the extra forward step is quite clearly seen, also the stepped sponsons.

Since taking this photograph I have fitted new and improved sponsons, reverting to my original non-stepped type, but using a swept back leading edge and a deeper and more buoyant trailing edge section. I now set these new sponsons at a greater angle of incidence. The new set-up has improved lateral stability on the water, and clean take-off is much improved because rougher water does not build up and sometimes get over the leading edge as it did in the case of the old sponsons.

The wing span of "Blue Goose," which was exhibited at the 1946 *Model Engineer* Exhibition, is 7 ft. 6 in., with a central chord of 14½ in. to the elliptical wings. The model has a delight-

Fig. 102. The author's 7 ft. span flying boat snapped whilst flying over Poole Water. It is powered by a German "Eisfeldt" diesel of 6 c.c.

Fig. 103. The 3.5 c.c. flying boat "Goose" shown, raised the author's original British flying boat record, and was designed for stability on and off rough open water at Poole. Note the long waterline to ensure the tail not being blown under when at rest in a wind.

Fig. 104. A large diesel-driven model of stable flying characteristics suitable for free flight or radio control known as "Bowden Whitewings."

fully steady flight over water and a very flat glide with a type of landing that is a pleasure to watch. I feel sure that more people will indulge in power flying-boats and seaplanes now that smaller and more portable models can be built with diesels like the little ones described earlier in this chapter.

In 1947 I made myself a medium-size flying-boat expressly designed for diesel engines of from 3.5 c.c. to 5 c.c. and capable of taking-off from the roughest possible water, also with the necessary stability to ride out rough water after landing. I decided upon these requirements because I had found so many days during our English summer weather when it was impossible to fly off open water such as Poole Water. The result can be seen in Fig. 103. This photograph makes clear the long water-line which prevents the tail being blown under when the model is at rest, and also the enormous sponsons which give unassailable lateral stability both at rest and when planing for the take-off. The great angle of the sponsons' leading edges ensures that quite large-scale waves do not climb on top of the sponson at take-off, which is the swiftest way to ruin a take-off. It will be observed that the model floats on top of the water and therefore the engine has not a great deal to do in raising the model on to its usual three steps. This model raised my original record for a while, powered with a B.M.P. of only 3.5 c.c. The duration of flight was curtailed by the small tank fitted to the engine. This model has led me to further and better designs for other engine capacities. The weight of boat " Goose " is 3 lb., wing span 4 ft. 8½ in. A new E.D. " Mark IV " diesel of 3.5 c.c. now powers this boat with increased vigour.

A Large General Purpose Flying Model

Fig. 105 is of interest in that it shows my large " Bowden Whitewings " starting on its take-off run powered by a 3.5 c.c. diesel. The tail has just come up and in a few more yards the model was airborne. The same model can be seen in Fig. 106 flying steadily around the camera, and also at rest on the ground in Fig. 104.

This model was heavily loaded with a coil and battery to

Fig. 105. The " Bowden Whitewings " large power model now produced in kit form, is here seen taking off. It is powered by a 3.5 c.c. diesel, or a 5 c.c. glow plug motor for radio control work.

Fig. 106. "Whitewings" airborne.

keep its normal balance for petrol engine flying as well as diesel. A $3\frac{1}{2}$-oz. clock timer was also fitted. The model has a large wing area, but the wing span has been kept reasonably short for a large model by giving the elliptical wing a very large chord of 16 in., the span being 6 ft. 6 in. This type of wing has the advantage of compactness and yet has the area of a normal parallel chord wing of about 7 ft. 6 in. It is also very stable laterally.

Those who are interested in what is known as the Reynolds' number and "scale effect" will realise that a large section is more effective than a small one of the same outline. Therefore the large chord is better as a weight carrier than a small chord. The only way to attain this with a short span is to use an elliptically shaped wing which keeps a low aspect ratio wing efficient. Hence the reason why so many of my models have this shaped wing outline.

There is not the slightest doubt that a 3.5 c.c. *petrol* engine of any make *would fail* to take this model off the ground under

Fig. 107. Mr. P. Smith designed this nice-looking 7 ft. span "Stentor 6," fitted with parallel chord wings. A number of these models have been fitted with radio after slight modification.

its own power and it would only just fly it. A 4.5 c.c. motor flies it comfortably in the air, but requires a bit of a push for R.O.G. work, but a 3.5 c.c. diesel makes the model rise off ground when loaded with all the spare electrical equipment of a petrol model. The evidence is in Fig. 105 and answers the question, does a good diesel produce more power than a good petrol engine?

The model "Whitewings" has had a great variety of petrol and diesel engines installed, because I designed it for a constructional kit of parts for the average aeromodeller and I wanted to make certain that it was capable of taking all sizes and power outputs of engines. The model is rock steady in the air and will fly in heavy winds that usually cause a paper or fabric covered plain longeron model of this size to retire due to damage. The fuselage is the sheet-balsa and fabric-covered type that I have already mentioned.

The prototype seen in the photograph is covered with nylon by the "wet system" of covering.

This model has proved itself a great weight carrier, and due to this and its great stability, I have flown various radio sets in this model.

A Fast Cabin Cruiser for Larger Diesels

Although a large boat is a nuisance to transport, there is something very attractive about such a craft on the water. It looks purposeful and can travel really fast over quite rough water

The hard-chined Vee-bottomed planing boat "Swordfish" gives a great thrill when travelling at speed. She is a little too fast for pond work unless throttled down, but is really first-class fun when let loose with a full tank of fuel from a dinghy on the sea. Fig. 108 shows the boat in action, but not flat out, because the waves were considerably larger than appear in the photograph on the day upon which the snapshot was taken.

This model has recently been fitted with an "E.D." three-valve radio receiver. Although the model is normally used for long runs on Poole Water, which has an alluring 140 square miles or so for this purpose, a special tankage system is being incorporated together with waterproofing of the radio so that if opportunity permits, the model will attempt to cross the English Channel controlled by radio by Taplin and myself from the deck of a full sized boat. This will require calm weather but should afford considerable fun, besides being a new model venture. This boat was originally built when I was stationed at Gibraltar before the last war, where it did much free running at speed in the harbour. I sometimes wondered what the ancient besiegers of this bastion would have thought if they could have seen into the future—a tiny boat speeding over the waters they had fought for beneath the towering old "Rock," leaving behind a thin trail of magic blue exhaust smoke and emitting a hitherto unheard of exhaust howl.

A number of engines have been tried in this boat. The hull planes quite well with a 5 c.c. diesel. If a larger diesel were on the market she would "take it," as she has run at great speed

Fig. 108. The author's large planing boat "Swordfish" travelling well in a rough sea. The model has since been installed with radio steering equipment.

Fig. 109. The "Swordfish" on half throttle runs into a choppy sea. The hull is 3 ft. 10 in. long.

powered by a 13 c.c. water-cooled petrol engine. A 9 c.c. glow-plug motor of the " hot " American type is most suitable too.

Fig. 110. An odd spectacle is afforded when the author's "flying plank " with reflex trailing edge wing, powered by a "Frog " 1 c.c. diesel takes the air. In calm weather these "planks " are very stable.

Fig. 110A. A powerful diesel installed in a small model heavily loaded with radio equipment is the ingredient for reliable stunt flying on rudder control alone. Here is Colonel Taplin running up his Mark IV E.D. diesel to test the radio against vibration prior to releasing the model to give a demonstration of stunt flying.

EXPERIMENTAL MODELS

Experiment is the spice of life to many of us in model work. The diesel makes an ideal engine for such work because it is self contained and the weight is concentrated. It can be clapped on to a model in a minimum of time, and give the maximum of flying time without disturbing engine adjustments, for observing the flying behaviour of one's brainstorm in the air or on the water.

The Swiss and German idea of a " flying plank " or rectangular wing without any stabiliser tailplane is an intriguing problem, and forms an unusual spectacle in the air. The secret is in the reflex trailing edge section which looks after longitudinal stability, whilst the fins at front and rear of the nacelle

Fig. 111. The Baigent speed control model has a very low frontal area with "E.R.E." diesel engine buried in the wing, driving the propeller of 8 in. diameter and 10 in. pitch, through an extension shaft. A cover snaps into position over the engine.

hold the model up in steep banked turns. These fins must be above and below the nacelle. I do not recommend the type in bad weather, although it flies well in calm air.

A SHAFT-DRIVE MODEL AEROPLANE

Fig. 111 shows a novel streamlined racing control-line model made by Mr. Baigent, housing one of his E.R.E. racing diesels of 2.4 c.c. The engine is buried in the wing with a small cooling slit. An extension shaft from engine to propeller is fitted. This provides a long streamlined nose with a thin circular fuselage giving a very low frontal area and the minimum of drag. The model is mounted on a dolly undercarriage which drops off when the aircraft becomes airborne. It lands upon a dorsal fin below the fuselage nose. Why are there not more speed models on these lines?

AUTOMATIC CONTROL BY PENDULUM USING A DIESEL ENGINE AS POWER UNIT

Mr. Norman, who is a sculptor of note as well as an outstanding aeromodeller with many ingenious ideas, flew a tiny scale " Typhoon " complete with cannon protruding from the wings in the 1948 " Bowden International Power Trophy." The tailplane elevator and dihedral of mainplane were all scale and unaltered. The model was powered by a " Frog 100 " diesel which would have normally turned over in the air such a small model with scale dihedral. Norman's model, however, took off perfectly, climbed steadily but fast like the prototype, and flew around under control as though a pilot who understood high-speed flight was at the controls. All this was done by simple pendulum controls to elevator and ailerons. Many people in the past have predicted failure of pendulum controls, saying that centrifugal force at turns would upset the controls. Norman proved otherwise, and what is more, he did it with absolutely simple pendulums. The elevator pendulum was mounted about the C.G. position in the fuselage under the bubble cockpit. It acted on the elevator like a control line handle. The ailerons had their pendulum mounted in the centre section of the wing, which was detachable.

Fig. 112. Mr. " Natznees " Norman flies a scale " Typhoon " under perfect control at the " Bowden Power Trophy," 1948, by pendulum control to scale elevators and ailerons. The model was powered by a " Frog 100 " diesel and flew as if a pilot who understood speed was at the controls. The ailerons and elevator sizes can be seen in the photograph of the model, which has a wing span of only 3 ft.

This eventful flight proved that free-flight scale models can be flown satisfactorily without any specially severe dihedral angle added if the controls are operated by pendulum. This model had a high wing loading and was fast, and completely confounded the theorists regarding centrifugal force dangers, which in theory should have occurred due to the high speed! Belgian competition models had previously used pendulum control to the rudder in 1947 in order to bring their high-climbing models out of a turn at the top of the climb.

A few details of the Norman " Typhoon " pendulum-controlled model will doubtless interest readers. The propeller is a three-blader made from fibre, and almost unbreakable. The blades are interchangeable, and bolted into the centre hub. The guns are mounted in rubber to avoid damage. Fibre wheel covers and fibre nose cowling are used. The nose with cowling is knock-off. The simple pendulum control has a patent pending, but there is no reason why any modeller should not design and make himself a satisfactory control for private use using simple lead pendulums, as has been done by Mr. Norman. The wing

Fig. 113. The author's control line flying boat, which successfully takes off and lands on water with the operator standing at the water's edge. A special forward fin is required to resist the inward pull of lines at take-off.

span of the model is 3 ft. Why should we not use aileron pendulum control on radio models, in order to prevent excessive bank on turns? I have used elevator pendulum control.

A CONTROL LINE FLYING-BOAT

It often happens that there is a beach or a pond for flying off water but not a suitable stretch of water for landing on after a free flight. Furthermore, there may not be a boat handy for recovery operations. A control line water-plane in these circumstances permits the modeller to enjoy most of the fun of flying off and over water. It further gives him all the thrill of taking off water and the landing, which are controlled by his skill. In fact, I find it by far the most interesting and intriguing type of control line flying, which tends to become dull after much repetition, for there are more hazards and difficulties to be overcome in water control line flying.

It sounds a simple matter just to attach control lines to a flying-boat or a float-plane and fly. I have found that the chances of this happening successfully are slight. The first flying-boat I tried as a control liner invariably pulled in on the lines at the take-off before the boat could get going at sufficient speed to plane. The left sponson submerged and the boat slewed round out of wind. I found the solution in the boat shown in Fig. 113. It will be noticed that the following points are incorporated.

(1) The hull is exceptionally long in front of the wing, so that a small three-ply fin offset to turn the boat outwards from the centre of the control circle at the beginning of the take-off run before planing speed has been attained, can be fitted well forward. This gives adequate leverage.

(2) The 3.5 c.c. "E.D." diesel is given a large offset outwards, and has plenty of spare power for the light weight of the boat, which sits on top of the water's surface.

(3) The planing surfaces are very generous and have *exceptionally large and wide sponsons.* The longitudinal flotation base is rather extreme and gives an effect like a tea tray which skids over the surface of the water, therefore facilitating take-off.

(4) A little down thrust is given. Up thrust prevents easy planing.

To fly, the operator stands on the bank of a pond, or, better still, a little way offshore in the sea off a sandy and easily-shelving beach. A helper releases the model from the shore with the model sitting on the water, facing sea or pondwards. The pilot then has a half-circle in which to get airborne.

For solo operation, I have rigged up a long wire prong which drives into the sand near the shore, with the top just above the water's surface. The wire has loops at the top through which a pin passes. A loop at the model's stern also goes around the pin. A line from the pin to the pilot allows him to start the tethered model in shallow water. He starts up his engine, walks to the control handle which is stuck into the sand on its prong, picks up the handle, wades into the water a few yards if possible,

Fig. 114. The Baigent 0.9 c.c. diesel model car is here seen behind the radiator cap of the author's 4½ litre Lagonda. Barely 1 c.c. makes friends with 4,900 c.c.!

for this gives a larger " half-circle," gives the pin a tug by the line which releases the model, and flies the boat off water, taking care to keep the boat low for the first half-lap after becoming airborne. The landing is great fun and requires a little skill in judgment should the engine cut over land. In this case, a " whip " of the wrist is required to carry the boat on to a flat glide, landing with a slight lowering of the tail just before touch down.

I personally use thin fishing line for the flying lines and give a dope of special fisherman's waterproofing mixture so that the lines do not become soggy and sink into the water at take-off time before the lines tighten.

MODEL DIESEL CARS

The diesel is suitable for the light-weight class of car.

The diesel engine is particularly suitable for the smaller types of model cars because, as in the case of aeroplanes and boats we can reduce weight by elimination of the electrical gear and its complication of wiring.

The diesel is cool running and therefore lends itself to good streamlining of the body work.

A LITTLE DIESEL RACE CAR

The 2 c.c. class of diesel is excellent for a beginner's race car of a very simple type, built up with a wooden chassis and balsa body. The power unit, less flywheel, will have an all-up weight of approximately 6 oz., and the excellent power output affords scope for the design of a medium speed model. The "E.D. Mark III" 2.4 c.c. diesel set up a record for its class with a speed of 41.7 m.p.h. in 1948.

The very simplicity of the diesel has produced some outstanding and unconventional designs in countries abroad, and in at least one case in this country. These designs have broken away from the more normal arrangement of the model car which has the engine driving front or rear axle *via* a longitudinal propeller shaft and a right-angled drive to the driving axle. Designs of this standard type are easily obtainable on the British and American markets.

A centrifugal clutch is generally fitted in the drive on model cars in this country, but this is often omitted in designs abroad. The diesel engine in the unconventional designs drives directly through the front or rear axle, thus doing away with the complication of right-angle drive. One of the fastest men in this

Fig. 115. Simplicity of design is the keynote of this Swedish car, the "Myran." It is powered by a 10 c.c. diesel, drive being on the front axle. *Photo* : "*Model Cars.*"

country is Mr. Cruickshank, who uses direct drive without clutch. He starts by pushing with a walking-stick. A direct drive diesel unit can be seen in Fig. 38, Chapter II.

A CAR FROM STOCKHOLM

The most elementary car is surely that from Stockholm called the " Myran " (Fig. 115). It is powered by a 10-c.c. diesel engine of " fixed-head " design. The drive is through the front axle, which is an extension of the crankshaft. The crank-case is then extended rearwards in the form of a streamlined shell to carry two wire extensions which form the rear axle, as in the case of the American " McCoy " Teardrop light alloy cast car, which created an American record for petrol cars.

A DIESEL CAR WITH A MULTI-PERSONALITY

Recently I examined Mr. Curwen's clever diesel car, having an engine with a detachable head which permits it to be run as a petrol motor, a diesel, or a glow-plug engine. The power from the diesel set up has proved superior to the petrol combination. The car also runs very well as a glow-plug motor.

This car has since become quite famous in the model car world and has demonstrated the high speed potentialities of the diesel engine as applied to racing cars. The powers of a diesel for car racing were much doubted by some people in the early days of diesel development. The car was run at the 1948 *Model Engineer* Exhibition.

Besides its multi-personality as a diesel-cum-petrol, cum-glow-plug motor, the engine has a dual mission in life, for it also drives a racing hydroplane. The best speed to date has been 24 m.p.h. on the water. The best official speed on land for the car is 62 m.p.h. These figures are most impressive for a capacity of only 5 c.c.

The engine has a bore of $\frac{3}{4}$ in. and a stroke of $\frac{11}{16}$ in. There are two exhaust ports and four transfer ports. Admission of the gas is by rotary disc valve. The cylinder head is fitted with a contra-piston which gives a variable compression ratio. It also has a sparking plug or a glow-plug. A contact-breaker is fitted which is used for starting from cold by spark ignition, for the engine has been found to be a difficult starter when cold, although perfectly normal when warm. The reason for this difficulty in starting from cold is the extremely high bore stroke

Fig. 116. A beautiful Delage diesel car by Mr. Baigent. The model is 13½ in. long and powered by an "E.R.E." 2.48 c.c. diesel unit with two centrifugal clutches in the rear brake drums.

ratio which produces a combustion space in the cylinder head having a high ratio of area to volume, and therefore high heat losses. Starting is done by using an external power pack which consists of a 3 volt dry booster battery and coil. Connection to the car-type contact-breaker being quickly made by a telephone jack. This method makes starting dead easy from cold, and is not needed when warm. Mr. Curwen has always been noted for his quick and reliable starting. His methods in dealing with a racing diesel are therefore interesting and also unique. The performance as a diesel is: Peak, 0.405 b.h.p. at 11,500 r.p.m., falling to 0.33 b.h.p. at 14,000 r.p.m. Maximum r.p.m. is approximately 18,000. Glow-plug performance is particularly good. The above figures are obtained from Mr. Curwen's well-known test apparatus and are not "guesstimation"!

I am always intrigued when I visit Mr. Baigent, of 10 Beverly Gardens, Bournemouth, who has set up a small works to produce model racing cars of scale appearance which actually do work at high speeds. Baigent uses his clever but very simple E.R.E. 2.48 c.c. diesel-cum-back-axle-unit (see Fig. 38, in the second chapter), containing two centrifugal clutches installed in dummy rear brake drums. This unit drives Maseratis, "Bugs," Delages, and other exciting cars at speeds around 60 m.p.h. Many "full-sized" car enthusiasts get their favourite car modelled by Baigent and use it around the pole as a working model.

In Fig. 116 we see a model of the late Dick Seaman's Delage with perfect wire wheels and tyres, with full cockpit equipment including spoked steering wheel and driving mirror, whilst in Fig. 118, we see a "special" with tubular chassis and independent front suspension with rear leaf springing that works, powered by a 1 c.c. diesel offset and on its side to fit into the coachwork seen in the next illustration. A back axle reduction

Fig. 117. Mr. Curwen s 5 c.c. diesel chassis. The engine is convertible to petrol or glow-plug by changing the cylinder head. The car has officially lapped at over 62 m.p.h. as a diesel, showing increased power over the petrol set-up.

Fig. 118. This " special " Baigent chassis is 11¾ in. long and powered by a 1 c.c. diesel geared 2 to 1 in the back axle. The engine is offset on its side to suit the body. Independent front suspension is fitted with leaf springs at the rear.

Fig. 118A. The " Special " seen in Fig. 118, has its realistic body in position. The exhaust " works."

gear is used on this model instead of the standard E.R.E. unit. The naked gearing seen in the centre of the chassis is merely to permit an offset engine and is 1 to 1. This car is a " runner " but is not meant for high speed racing as in the case of normal Baigent racers. It is 11¾ in. long.

PLANS AND KITS OF MODELS FOR DIESEL ENGINES

I hope that the foregoing descriptions of different sizes and types of model aircraft and boats will give the reader a very fair idea of the capabilities of diesel engines as we know them to-day. I also hope that the data given will inspire many people, trembling on the brink, to make the plunge into designing and constructing their own diesel craft. There is a great deal of fun, scheming and occupation to be obtained, not to mention the thrill of seeing the resulting models perform in the air or on the water. Experimental work can never be dull.

For those who like to start off with designs, either in kit form or plans, some of my models shown in this book have been produced either as kits or as plans. These can be obtained from B.M. Models, 43, Westover Road, Bournemouth. The boats " Sea Swallow," " Flying Fish " and " Swordfish " are obtainable in plan form. The " Flying Fish " is also produced in made-up hull form with suitable transmission, propeller and flywheel. The little " Meteorite " aeroplane for baby diesel engines, and the large model " Whitewings " are obtainable as kits or plans. The " Humming Bird " and the little 36 in. flying-boat " Wee-Sea-Bee " as plans. The control line model " Bullet " and the monocoque free flight " Satellite " are now available as plans or kits.

Since I wrote the first edition of this book in the early days of diesel development, a full range of excellent plans and kits to suit diesel engines has been placed on the market by all the well-known manufacturers, and suitable plans for construction have been published by the model press. A survey of these would fill a further book. A glance through the advertisements in the model journals will whet the appetite of all enthusiastic modellers or make enthusiasts of modellers !

Fig. 119. Mr. Fjellstrom's car with cowling removed to show the mechanism. The engine, with a capacity of only 0.25 c.c. drives the front wheels through double helical reduction gearing.

RADIO CONTROL AND THE DIESEL

Radio control is the ultimate in model flying and boat work for most modellers, and now that it is not necessary to have a licence provided a permitted wavelength is adhered to, radio is gaining many adherents. The early commercial sets have gone through their period of development growing pains, in which the satisfactory ones have proved themselves in the hands of the user. The diesel (and glow-plug engine) forms a particularly useful motor for radio control models of the small to medium size. For instance, the new baby radio sets can be flown by the more powerful of the smaller diesels in little models of around 45 in. wingspan such as my little " Meteorite " and " Radio Brumas " shown in this chapter, or the larger sets can be operated with engines ranging from the powerful 3.5 c.c. " E.D." diesel to the selection of 5 c.c. diesels described in Chapter I. One would like to see some well-known engine manufacturer produce a 10 c.c. diesel as has been done in Italy with success, for the larger radio model which will always fly more convincingly than a small model. The diesel is so very convenient for radio work as it saves extra battery, coil, etc., weight for ignition, which added to radio batteries may mount up alarmingly when spark ignition motors are used. Furthermore, there is quite sufficient electrical equipment to cope with in the radio itself without the extra electrics of the petrol engine.

It is quite simple to get two-speed control with a diesel if desired, by a simple throttle control.

A short explanation of radio equipment may introduce this most exciting and ever alluring form of operating diesel engined models. I am convinced that once the reader has got successfully into radio control he will seldom operate any model without it! The trouble has been, that there were too many early failures amongst enthusiastic modellers when using certain tricky and unreliable sets fitted into unsuitable models with overpowered engines. Many people gave the game up, which did harm to the good name of radio. None of this need now occur. Having tried every well-known radio set on the market, I now know the most suitable for my own requirements, which, like those of most modellers who are not essentially radio fans, are :—(1) To tune the receiver quickly by one simple operation. (2) Rely upon flying or boating without further " fiddling " with the radio gear. (3) Knowing that the radio can not cause a crash through the old bugbear of a stuck on " relay " and therefore rudder control.

I now have six of my models rigged up for radio flying, and one speedboat, and hope shortly to be sailing my model yacht by radio. All my gear can now be relied upon to comply with the requirements stated above, with reasonable luck, but I have thrown away quite a lot of the early radio equipment in disgust !

THE USE OF RUDDER ONLY

Many modellers contemplating radio control for the first time imagine that the actual flying is merely a matter of turning the model by rudder from left to right, or making it climb or dive by giving the machine up or down elevator. I strongly advise the newcomer to start off with the rudder only, for even this will tax his flying control ingenuity, and he should remember it is possible to do up to loops with rudder control alone. The leading Americans, who started radio long before we did, have won their National contests for many years past with simple rudder control. Doubtless we are coming to the time when the " experts " use a greater number of controls, but how often does one yet see such a one flying with more than rudder and, perhaps, engine control. There is one proportional control that I know is working satisfactorily. Model radio has so far been in sequence of controls, and not selective, by which I mean that left rudder is followed by centralise and then right, and one must go quickly through right if further left is desired. Proportional control means that half left or right rudder can be given, followed by full rudder, or full rudder skipped rapidly through. It is quite hard enough to remember which rudder you gave last time when an emergency occurs and the model is flying away from you in the far distance, without getting mixed up with going through elevator controls to get at opposite rudder. My advice to the newcomer is to " take it easy " and start off with rudder, being sure that you arrange your rudder *so that it holds on whilst you hold the signal button down, and automatically returns to neutral when you release.* Any other system will cause a crash if the next signal does not arrive, as I will explain in the next paragraphs.

A STUCK ON RUDDER CAN CAUSE A SERIOUS CRASH

Modellers who have not tried radio forget that a full-size aeroplane and model, however automatically stable, will increase its bank, eventually getting into a spiral dive of increasing intensity, if the rudder is kept hard over for more than a short time. Should the radio stick on, due to the relay failing to " come off " the rudder will remain hard over and the model will increase its bank until it is eventually on its side in the air. The nose will drop because the rudder now becomes an elevator in the *down* position. A crash usually results. *Therefore we must not keep turning too long and we must not have a radio set that is prone to " stick on."* This has been the besetting sin of fliers and radio sets in the early days, and caused many a stable model to end up as matchwood, to the amazement of its owner who had flown it free-flight without damage. It also means a well designed model and a perfectly reliable radio set. One well-known model writer maintained that almost any model could be fitted with radio and fly well. I could not disagree with any man more, and most of the leading American radio men support my view that a radio model must be stable, so that when you have got it into an odd position in the air through your playing with controls, if you centralise your rudder, the model will *quickly* iron out its unstable attitude by rapid recovery of its own. Except for stunt work a pylon model is not suitable, for the thrust line is too low in relation to the centre of lift and drag. This gives great variations of trim as the model is turned from left to right. The best set up is either a high wing with engine mounted so that the thrust line is not far below the wing, or a low wing with plenty of dihedral. It is very easy to zoom a radio model all over the sky and call it stunting, but very hard to fly one with real precision around a course and then do a spot landing. In my opinion the latter type of radio flying should be encouraged.

In short notes such as these it is obviously not possible to go into the movements of flying or the radio itself in detail, but I hope I can give the newcomer some fundamental facts that will save him a lot of wasted money and much disappointment.

Fig. 120. The author's "Whitewings" is turning to the left under radio control. The nose is held up by area low down due to spats and filled in undercarriage legs, which act as forward keel.

FLYING THE MODEL

For example, if we start turning the model to the left, bank will automatically increase as the turn is persisted in, for there is no pilot to hold off bank by opposite aileron or to give some elevator in a tight turn. The model will begin to lose height rapidly with nose down, so in good time we take off our signal, for it takes time for the model's nose to come up with its increased speed as the machine straightens out. As soon as its nose is up after the resulting shallow dive we can give a very quick flick to the right rudder because it will be remembered that earlier I explained we have to go through controls in sequence. We can now restart our left turn if we want to bring the model right round, and we can turn in these " instalments " remembering *that the engine torque reaction will always assist a left turn and resist a right turn*, and that on the glide we turn without torque. We therefore have given our motor a right offset to

Fig. 121. The author's large radio model " Poole Puffin " with deep underbelly for stable turns, is here seen ruddering into wind after take-off, by a series of short sharp movements of the control button

make it eliminate torque effect as much as possible and give as straight flight as possible under power with rudder centralised. We have also made certain that the engine is not too powerful for the model and that a little " down thrust " is built in to prevent stalling under power as the model is turned from left to right. Far too many newcomers to radio fit an engine of too great power and find themselves all over the sky more out of control than in it ! They forget that *in a free flight model they may have prevented undue climb and stalling under power by making an overpowered model hold its nose down by turning in circles.* Others who have very powerful motors try to control this by fitting a rudder with very minute movement for radio. This does not work out in practice, for such a rudder control will turn the model well to the left without losing much height, but if there is a cross wind when trying to turn to the right against torque the model will merely fly straight ahead out of the ken of the operator.

On the other hand a reasonably powerful motor is required

Fig. 122. Mr. Honnest-Redlich, the designer of the E.D. " modulated " radio set, tunes in his receiver before giving a demonstration of radio flying to the Dutch Forces.

to gain height after turns. I can not stress too strongly that a correctly powered engine is required with the right amount of offset and down thrust, and with the thrust line passing not far below the centre of lift and drag of the wing.

From what has been said, it will be evident that it is dangerous to turn for more than a short way just after take-off, for if a mistake is made and too much height lost with a subsequent dive at speed on the turn, the model may not have time to recover before it hits the deck in that most devestating of all manoeuvres, a cart wheel crash. Therefore, keep the turns very short and sweet until height is gained. (Fig. 121, to remind the reader.) The reader will also probably realise that if he wants to lose height when the model is climbing too fast he merely spirals down in a tight turn, but the wings must be robust and very well fixed, for there are big stresses in this manoeuvre, as the model comes out with a mighty zoom, which can be checked

Fig. 123. Colonel Taplin, a well-known radio flyer, sends off his "Radio Queen" for an impressive demonstration flight at Manston aerodrome. The engine is an E.D. diesel and the receiver a "modulated" three valve E.D.

by flicking it over for a very short turn of rudder as the zoom is half completed. If great speed has been gained, it is possible to loop the model if it is left to recover on its own. This is

Fig. 124. The author's layout has a forward battery compartment with receiver and servo behind. The receiver is held in position by rubber bands to four cross members in the fuselage. The white tuning arm on the receiver is reached by forefinger through opening in fuselage side. Below are mainswitch and tuning socket for headphones, or meter.

how we stunt when desired. It is best to use an overpowered machine with clipped wings for stunting.

STABILITY FEATURES

For quiet normal flying, a deep bellied fuselage or an undercarriage with legs well filled in to form area low down forward like the keel of a boat, will help to hold up the nose on turns. It will be observed that most of my radio models shown in this book, have a deep belly. Where this is not done, I provide the desirable keel area, as seen in my more normal looking " Whitewings," by fitting large wheels or spats allied to longer filled in undercarriage legs. These two methods can be seen in Fig. 120 which shows my " Whitewings " being turned nose up under radio, whilst Fig. 121 shows my large " Poole Puffin " which sports a large underbelly reminiscent of a bird with a belly full of fish. The well-known American radio modeller,

Dick Schumaker, admits that he has fitted under fins to the belly of some of his leaner looking models to gain this desirable holding up the nose on turns.

THE NORMAL COMMERCIAL RADIO SET TODAY

Most radio sets have a similar general set up. This takes the form of a transmitter on the ground with a high tension and low tension "wireless" battery bought at any radio shop. The L.T. is sometimes in the form of an accumulator of two volts, which can be trickle charged, as in the case of the "E.D." set. A receiver and servo motor with their batteries are installed in the model. The receiver detects the signal, and switches on and off the servo motor by the aid of a "relay." The servo motor operates the rudder or other controls and has a clockwork motor or twisted rubber motor to do this. The receiver has H.T. and L.T. dry batteries, of the deaf aid type or flash lamp, and the servo motor has a flash lamp battery. Some call the servo motor an actuator. See Figs. 124, 125, 126 and 127, which explain pictorially the general set up.

THE TWO MOST USUAL PRINCIPLES UPON WHICH SETS WORK

(1) The first principle of "dipping" the current is used by several of the older sets and many of the modern baby sets using the new mini-valve. With radio switched on, but receiving no signal, a "standing current" is registered in the valve of the receiver, and the "relay" arm is therefore attracted down or "in." This "in" position switches off the servo motor's battery. The servo is therefore not operating the rudder or other control, i.e., the "relay" contact points positioned at the other end of the relay arm are *open*. The arm is pivoted at its centre. On receipt of a signal from the transmitter the current in the receiver valve is reduced or "dipped." This current drop can be read by a plugged in meter, The "dip" at best is very small, being measured in milliamps (one-thousandth of an amp.).

As the current is reduced or "dipped" the attraction to the relay arm dies and the arm is released, thus closing the points

Fig. 125. The fuselage under construction shows the four cross dowels upon which the receiver is slung. Servo is behind the wing platform

at the other end by spring tension. This closing of the points makes a circuit for the servo battery which switches on the servo, which in turn pulls over the rudder. To operate against air or water pressure, the servo finds power in a clockwork spring or a twisted rubber motor.

The relay is therefore released as long as the signal is on and the current dipped. As soon as the signal is stopped the full standing current again passes through the valve, and down goes the relay arm, thus opening the points which switch off the servo. There are various variations of this theme, but that is the main principle.

It will be realised that with this system, *if the battery voltage drops, the current may be insufficient to open the points by attracting the arm down again*, in which case the rudder will stick on hard over and the disastrous spiral nose dive will most likely end the model's days. This is the weakness of the principle, and precautions must be taken, by fitting a good relay and keeping

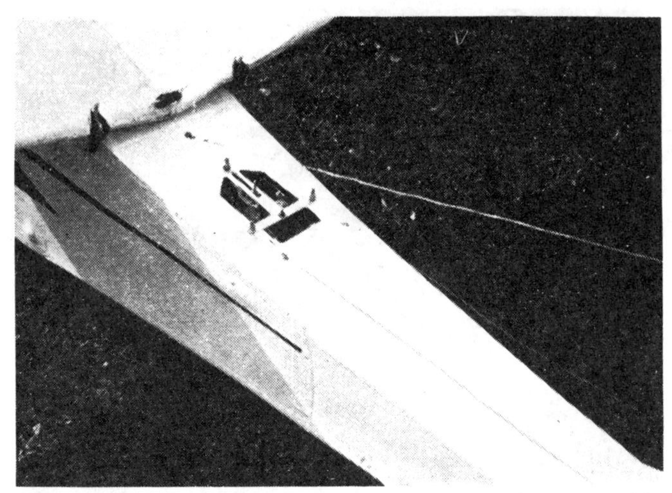

Fig. 126. The servo is as near the centre of gravity as possible with fishing lines to rudder. Wires would upset the reception. The aerial is led to the fin tip and attached by a rubber band to a wire hook.

battery voltage up, and also signals from the transmitter powerful. This system is cheap to construct and the wiring very simple. The latest mini-valve made by Hivac which is a gas filled " thyratron " valve has a lower battery consumption and a larger " dip " than its larger predecessors of the early days, which, together with good manufacturing methods of relay and care by the user that his transmitter is well up to scratch, and the batteries in the model are " up," all help to obtain reliable operation. Unfortunately all these points were not obtained in many of the early " dipping " sets, which gave the radio control game a bad name for crashes of valuable models due to sticking on controls. It must, however, be clearly recognised that the " dipping " principle, however carefully looked after, can never be quite as " safe " as the principle described below. However, owing to its light weight with the new mini-valve and its lower battery consumption and therefore smaller batteries, the " dipping " principle is very suitable for small models, or boats where an occasional on control does not spell trouble.

(2) *The second principle of " rising current "* :—as used by the well tried three valve " E.D." receiver with its special transmitter giving off a " carrier," works on an exactly opposite system. On receipt of the signal the current in the receiver is raised instead of being dipped. This principle has various undeniable advantages which preclude danger of stuck on relay and therefore rudder, but it costs a little more to produce and works out a trifle heavier as regards receiver weight. For instance, the E.C.C. and the E.D. firms make baby and cheaper but very good " dipping " sets which all up weigh approx. 7½ oz. These suit the baby model admirably. The three valve receiver and batteries weigh just over double this load. In the future this will doubtless be reduced when the mini-valves and their smaller batteries are ultimately employed for the three valve " raised current " system. This safer system should then be sufficiently light for even the baby models, although it will still doubtless cost more. From much personal experience of all the past sets of all makes I have come to use the three valve " rising current " set in all my models over 5 ft. wing span. Naturally, I use the little " dipper " for my baby models around 45 in. span for weight considerations.

This three valve " rising current " " E.D." set gives utter simplicity of tuning, virtual immunity from stuck on controls, and a very outstanding range due to the three valves, etc. Range is very desirable, and was very poor on the early " dipping " sets. With the " E.D." " raised current " set one can fly the model in range as far as, and farther, than it can be seen at ground level. Range at height is always even greater. With this principle the worst that can happen is a flyaway under normal centralised rudder stable free flight conditions, should batteries drop or a signal fail to be received.

Let us follow through the operating sequence. A high frequency " carrier " is permanently radiating from the transmitter as long as it is switched on. (Incidentally, this means that such a transmitter *can not* operate a " dipping " receiver. For their little " dipping " receiver the E.D. firm provided the

Fig. 127. The complete set up can be seen in this photograph. The rudder tab is attached to the fin by fabric hinges like the elevator of a control line model.

first powerful commercial 4 watt transmitter which makes for reliability. In the past the "dipping" sets suffered from low power of approximately 1 watt in Britain, whereas the Americans all used about 4 watts.) E.C.C. now use high input.

With the "rising current" principle the "carrier" is "modulated" on the transmitting of a signal. This modulation can be heard through headphones plugged temporarily into the model for tuning. In fact this is the only tuning necessary. The operator slightly moves a small tuning arm (which can be seen in Fig. 124) until he hears the loudest note in his phones, whilst an assistant holds on a signal by press button. The set is then ready for operation. It is as simple as that! It is, therefore, suitable for a beginner or someone who enjoys flying rather than radio adjustment. A small boy can tune this set.

This modulation of "carrier" allows a greater current to pass through the valves. *The increased current operates the relay.* Our old friend the "standing current" is kept low before signal to approximately only ½ Ma (milliamp). This is done by the holding down of the current by a 6-volt grid-bias battery, which has no drain upon it other than loss of energy through old age. The grid-bias therefore need not often be changed, but it is a very important battery, and must be up to its six volts. The " relay " is set at the works to click in at 2 mA. *This setting should never be touched by the owner*, a very good feature for the beginner. As the signal comes in and the current rises, it goes up to between 3.4 and 4 mA, and on its way it passes the point of 2 mA at which the relay operates the servo motor, and so the rudder.

It will be noted that there is a very wide tolerance in this principle, should the batteries drop in voltage or the receiver be a bit off tune. The only possible danger of a sticking on relay could occur if the grid-bias battery (*which has no drain*) becomes low due to neglect, when the " standing current " will obviously rise too high and perhaps pass the magic 2 mA at which the relay clicks in. But this could only happen through gross carelessness, and need not be considered seriously, provided the operator understands the principle. For instance, if he were to fit a 4-volt battery in error this would not hold down the standing current sufficiently low. It is the only system, where I definitely know that if my model spirals in, it is due to my own fault of operation or design of the model.

Perhaps I should mention here that in all types of sets, whatever principle is used, residual magnetism can form at the relay and hold it on, causing a stuck on control. This often happened on some of the earlier sets, but today modern manufacturing methods of tinning the relay arm and using suitable materials, should have eliminated this defect, when a set is bought from a reputable manufacturer. We can therefore dismiss the trouble. Naturally a " dud " relay will cause a stick on.

Purely as a matter of interest, and not of necessity, it is possible to watch the " E.D." three valve receiver in action by plugging in a milliampmeter to the headphones socket (normal tuning can be done entirely by headphones) when the alteration of current can be observed as follows.

(a) With transmitter switched off, the meter will show approximately 1 mA " idling current."

(b) With transmitter switched on but no signal sending, the "standing current" will read approximately ½ mA.

(c) As signal is sent from transmitter, the meter will show the current rise up to 3.5 to 4 mA. On the way, as this passes the point at 2 mA, the relay will be heard to click in and the servo seen to operate.